T0140228

Advanced Sciences and Technologies for Security Applications

Series Editor

Anthony J. Masys, Associate Professor, Director of Global Disaster Management, Humanitarian Assistance and Homeland Security, University of South Florida, Tampa, USA

Advisory Editors

Gisela Bichler, California State University, San Bernardino, CA, USA

Thirimachos Bourlai, Lane Department of Computer Science and Electrical Engineering, Multispectral Imagery Lab (MILab), West Virginia University, Morgantown, WV, USA

Chris Johnson, University of Glasgow, Glasgow, UK

Panagiotis Karampelas, Hellenic Air Force Academy, Attica, Greece

Christian Leuprecht, Royal Military College of Canada, Kingston, ON, Canada

Edward C. Morse, University of California, Berkeley, CA, USA

David Skillicorn, Queen's University, Kingston, ON, Canada

Yoshiki Yamagata, National Institute for Environmental Studies, Tsukuba, Ibaraki, Japan

Indexed by SCOPUS

The series Advanced Sciences and Technologies for Security Applications comprises interdisciplinary research covering the theory, foundations and domain-specific topics pertaining to security. Publications within the series are peer-reviewed monographs and edited works in the areas of:

- biological and chemical threat recognition and detection (e.g., biosensors, aerosols, forensics)
- crisis and disaster management
- terrorism
- cyber security and secure information systems (e.g., encryption, optical and photonic systems)
- traditional and non-traditional security
- energy, food and resource security
- economic security and securitization (including associated infrastructures)
- transnational crime
- human security and health security
- social, political and psychological aspects of security
- recognition and identification (e.g., optical imaging, biometrics, authentication and verification)
- smart surveillance systems
- applications of theoretical frameworks and methodologies (e.g., grounded theory, complexity, network sciences, modelling and simulation)

Together, the high-quality contributions to this series provide a cross-disciplinary overview of forefront research endeavours aiming to make the world a safer place.

The editors encourage prospective authors to correspond with them in advance of submitting a manuscript. Submission of manuscripts should be made to the Editor-in-Chief or one of the Editors.

More information about this series at https://link.springer.com/bookseries/5540

Reza Montasari · Hamid Jahankhani
Editors

Artificial Intelligence in Cyber Security: Impact and Implications

Security Challenges, Technical and Ethical Issues, Forensic Investigative Challenges

 Springer

Editors
Reza Montasari
Hillary Rodham Clinton School of Law
Swansea University
Swansea, UK

Hamid Jahankhani
Northumbria University
London, UK

ISSN 1613-5113   ISSN 2363-9466 (electronic)
Advanced Sciences and Technologies for Security Applications
ISBN 978-3-030-88042-2   ISBN 978-3-030-88040-8 (eBook)
https://doi.org/10.1007/978-3-030-88040-8

This Springer imprint is published by the registered company Springer Nature Switzerland AG
The registered company address is: Gewerbestrasse 11, 6330 Cham, Switzerland

Contents

Handling Novel Mobile Malware Attacks with Optimised Machine Learning Based Detection and Classification Models

Ali Batouche and Hamid Jahankhani

Abstract Malicious behaviour analysis is one of the biggest and most prevalent challenges in cybersecurity. With the dominance of the Android ecosystem, a significant number of frameworks were proposed to address the huge number of malicious attacks targeting the consumer base of this platform. Although still developing, the application of machine learning techniques for malware detection has recently experienced a growing interest due to their potential to achieve better results compared to traditional techniques. However, the effectiveness of detection by learning varies according to the used features and models. Moreover, its application to mobile malware detection is even more challenging given the deployment constraints. In this paper, mobile malware detection is cast as a classification problem, and four main and relevant questions are considered: (1) Which set of features is more relevant for effective detection using ML models. (2) which models are best performing for this type of tasks (3) which solution can be the most lightweight and most effective for real-time detection (4) And finally, how can these models be optimized to address the risk of a zero-day attack. This paper describes a comprehensive investigation of the potential of traditional and advanced ML models to address the aforementioned issues. As a result of this in-depth study, a testbed has been prepared using 168 different models and three recent datasets. Furthermore, the main contribution of this work lies in the development of novel models that outperformed the state-of-the-art proposed approaches. One of which, combined early integration with Extra Trees Classifier which achieved a detection rate of 99.94% and an AUC score of 99.91%. Further experimentations were conducted on the deployability aspect of these models, where results have shown that Boosted algorithms offered the best balance of detection rates and resource utilisation for a lightweight and robust malware detection solution. Furthermore, this comprehensive analysis helped in one hand, gain more insight into the role of features in the learning task, which led to the identification of a set of characteristics that we believe should be considered to develop an effective dataset to counter novel malware attacks. On the other hand, it helped in highlighting future developments and the missing components in this field, where we ultimately

A. Batouche · H. Jahankhani (✉)
Northumbria University, London, UK
e-mail: Hamid.jahankhani@northumbria.ac.uk

© The Author(s), under exclusive license to Springer Nature Switzerland AG 2021
R. Montasari and H. Jahankhani (eds.), *Artificial Intelligence in Cyber Security: Impact and Implications*, Advanced Sciences and Technologies for Security Applications,
https://doi.org/10.1007/978-3-030-88040-8_1

proposed a framework that builds on this analysis to provide a better approach for future studies.

Keywords Android · Malware detection · Machine learning · Ensemble learning · Deep learning · Fusion · Lightweight · Zero-day attack

1 Introduction

Despite Huawei's ban from the Android market in 2019 and the launch of their own Harmony OS as their new operating system, the Android platform still had an 84.8% market share in 2020 and is projected to hit 85.5% by 2022 [1]. The android platform did not stop growing and is actively expanding to other platforms. At the moment of writing this paper, Android is already available on multiple platforms ranging from smartwatches to car infotainment systems. With all of this development, malicious attacks against this platform became increasingly widespread, with new attack vectors emerging every year. Several security agencies have reported this spike of malicious attacks in 2020, but the decline in the detection of new malware attacks has intrigued researchers' interest [2], as this implies that conventional and currently deployed Intrusion detection Systems (IDS) systems are rendering inefficient against the emerging attack techniques. Hence there is a big emphasis on the study of zero-day malware detection in literature, developing new techniques and approaches to enhance the detection efficiency.

One developed approach takes advantage of Machine Learning (ML) and its power of generalisation. This approach revealed better potential compared to the conventional methods coping with the limitations introduced by the traditional signature based classification, which was inefficient in handling new unrecorded malware categories. The primary objective of using machine learning techniques is to facilitate the discovery of patterns and the extraction of information from large collections of data. Machine learning can be used to achieve optimal detection rates once enough training data is available [3]. The most recent methods make use of Deep Learning, a branch of machine learning that can yield better outcomes [4]. As opposed to conventional machine learning techniques, deep learning approaches require more extensive datasets but can achieve higher detection rates. With the power of learning from example, researchers took different approaches to handle this classification task. As would be explained in later sections. Three main directions are being adopted recently, namely static, dynamic and hybrid methods, depending on the analysis type and extracted features.

Furthermore, since this work targets mobile systems, many constraints should be taken into consideration regarding battery, processing power and memory devices. Therefore, a lightweight application is to be considered to fit the purpose. The nature of these systems requires real-time detection in order to be as effective and as adequate as possible in response to the increased attempts against these devices that became a critical asset for every person's business. Thus, a trade-off arises between performance and efficiency if the system is to be deployed on the host machine itself.

So with that being said, an interesting question arises: How can we leverage the advanced machine learning te66chniques to increase the detection efficiency in terms of accuracy, false alarm rates and zero-day robustness in mobile devices, given the current developments of new malware datasets?

To investigate this matter, the rest of this paper is organised according to data analytics life cycle project. Sections are outlined as follows.

Section 2 is aimed at revealing the research gaps and identifying the objectives of this paper. That includes an in-depth review of the latest proposed solutions. Four main objectives were extracted as a result.

2 Related Work and Research Gaps

To properly frame the problem and set the initial hypotheses, the following will explore the literature and the proposed solutions and datasets. One key point to consider is that the model we are aiming to achieve should be trained on a bench-marked dataset containing enough samples and features, so that the model is capable of handling evasion techniques, while maintaining consistency to recognise and identify new attack vectors.

2.1 Discovery Phase and Problem Understanding

To properly frame the problem and set the initial hypotheses, the following will explore the literature and the proposed solutions and datasets. One key point to consider is that the model we are aiming to achieve should be trained on a bench-marked dataset containing enough samples and features, so that the model is capable of handling evasion techniques, while maintaining consistency to recognise and identify new attack vectors.

Related Work

Several approaches have been presented, each with its own strengths and weaknesses. The following literature review would reflect on several significant studies that took an interesting approach and achieved promising results.

Mas'ud et al. [5] has conducted a series of tests on their lab-built dataset, consisting of system calls recorded from 30 benign apps from the google store and another 30 malicious apps from MalGenome project using Strace. Their testbed combined several sets of feature selection methods (including the Chi-square method and others inspired by other studies) with different ML classifiers (NB, KNN, J48, MLP, RF). The evaluation was based on Accuracy, True Positive Rate and False Positive Rate. Although the data collection was obtained by running the application on real devices, it can be argued that the number of samples is not significant enough to get an accurate reading, nor the nature of collected features which only consisted of one single type of features. Best results were achieved using the MLP classifier when combined with extracted features from chi-square algorithm; Accuracy 83%, TPR 90%, FPR 23%. This study, however, sheds light on how high dimensionality and feature correlation can have a negative impact not only on the performance of the classifier but also on the final outcome of the trained model.

Other efforts using machine learning algorithms was conducted by Ihab Shhadat et al. [6], however, this study focused on the behaviour of the application. The experiments relied on a dataset composed of 984 malware and 172 benign applications, where behaviours such as network traffic, memory dumps, file tracking and system calls were tracked and recorded. Results differed between binary and multiclass classification, however, RF, HV and DT algorithms have consistently achieved better performance than other classifiers. The difference in this study is that feature reduction has indeed increased accuracy but when other metrics were measured (such as the recall and precision), a decrease was reported in this case. The author has asserted that this may be due to the imbalance in the families of the dataset as malicious applications compose about 85% of the entire dataset. Hence it is essential to balance the dataset in our experiments to avoid the occurrence of biased results.

Arshad et al. [7] proposed SAMADroid, a three-layered hybrid analysis model for android operating systems. In the first layer, data extraction was conducted using both static and dynamic analysis. The second layer is a combination of local and remote hosts. Dynamic analysis is conducted on local hosts using Monkey Runner. The Remote server analyses the behaviour of the application generated by the logs and extracted static features. Finally, the last layer, also taking place in the remote server, performs the classification task based on the data extracted from the previous layers. This separation is very beneficial looking at the nature of mobile devices, especially as the dataset is stored in the memory of the server, hence reducing the memory and calculation load off the mobile device. However, practically speaking, such method will raise considerations towards the confidentiality and integrity of the uploaded data itself. The classification task in SAMADROID uses the SVM algorithm for both dynamic and static features, classifying the application into three categories: "Legitimate", "Malware", and "Risky" if there is a contradiction in the results of the classification. For evaluation purposes, the authors used the Derbin dataset -a rich but relatively old dataset, to measure their model's performance based on Accuracy, True Positive Rate and False Positive Rate. When it comes to classification, they have achieved the best scores using Random forest with static features (99.07%) and RF/SVM using dynamic features (82.76%).

Since this proposed approach focuses on the lightweight aspect, performance benchmarks were conducted to compare the overhead created by SAMADroid in contrast to other commercial tools (including; Avast Mobile Security, AVG AntiVirus, Avira, Kaspersky Internet Security, McAfee Security and 360 Security). Results showcased how SAMADroid was the most lightweight solution in memory allocation, power consumption, and performance overhead.

Yoo et al. [8] Proposed "AI-Hydra" a machine learning-based model that combines Random Forest with a deep learning model for Binary Classification. The proposed model also uses a 3-stage process: First a feature Extraction stage, that extracts five feature classes from static analysis and another 5 classes from dynamic analysis, using a dataset consisting of 6,395 (4,160 malicious and 2,235 benign) labelled files. Second, the classification stage generates results via four classification models, including two Deep learning models and two Random forest models, where one uses static features, and the other uses dynamic features. Lastly, a final decision-making stage would take the output of the models, and through a rule-based majority voting scheme the category of the file would be determined. The model reached a true positive rate of 85.7 and 16.1% for False positive rate, and has averagely spent 60.9 s per file using Intel XeonCore Silver 4116 CPU 2.10 GHz with 48 cores, 128 GB of memory, and an Nvidia Titan V. The proposed solution proved efficient in their test bed, under which the model competed with other commercial solutions. However, the proposed methodology is a heavy computational burden since multiple Machine Learning and Deep Learning models need to be loaded at the same time in order to make a decision. Also, the heavy 2 stage pre-classification could also delay the entire procedure, particularly as the number of features grows in practical circumstances. Further, there is still potential for improvement in the obtained results, particularly when state of the art algorithms produces better results by using less resources. Thus, as a lightweight option for mobile devices, such a strategy may not be convenient.

Authors in [9] proposed the DL Droid framework for android malware detection. They created a dataset containing 419 static and dynamic features observed from over 30,000 applications running on real devices. This work further explored the stateful versus the stateless input generation using Monkey and DroidBot respectively, in order to test whether the type of input would affect the performance of the trained model. In fact, they proposed a deep learning architecture using 3 hidden layers with 200 neurons each. The results surpassed other machine learning algorithms results as they achieved 95.4% accuracy rate with stateless input dataset and 98.5% on stateful input. This work shed light on the importance of the quality of the dataset and also ranked the recorded features based on InfoGain, which is why this project considers this dataset as a good benchmark for the proposed models.

Another vector of classification using deep learning takes advantage of text and image classification. Such works include the experiments conducted by Nan Zhang, et al. [10], where they applied natural language processing methods to mine the difference between malicious and legitimate applications. Their approach takes advantage of TextCNN which is a layered model composed of (1) embedding layer, (2) convolutional layer, (3) max-pooling layer, (4) fully connected layer, and (5) softmax layer. After generating reports of apk files in JSON format, they extracted four static features: Permissions, Services, Intents, Receiver. Once features are selected, the TextCNN model is used for the classification task. To evaluate their approach, four built-in datasets were used, containing malware and benign apps from several resources. Metrics including: Accuracy, precision, Recall and F1-score were used to evaluate the detection performance. The results showed that TC Droid outperformed other proposed methods. Which shows how efficient malware detection using NLP can be a practical approach while removing the overhead caused by feature selection.

Other experiments have been conducted in the same field of text-based classification [11] proposed Amalnet a DL framework based on Graph convolutional networks (GCN). In this research the authors have applied GCN to model graphical semantics in combination with IndRnn (Independently Recurrent Neural Network) to extract semantic information into several datasets based on different representations (character, word and lexical features). Their experiments showed how efficient this technique could be according to the dataset used and family distribution, features selected, zero-day sustainability and even runtime performance.

CNNs are also used for image-based classification tasks [12] proposed a framework to detect malicious attacks in Industrial Internet of Things (MD-IIOT) based on colored image visualisation using DCNN (Deep convolutional neural network). Two datasets were used for evaluation and benchmarking this approach: Leopard Mobile dataset and Malimg. They measured the impact of different image ratios, achieving best results at 229 × 229 ratio with less false positive rate (98.79% overall testing accuracy in Malimg).This study shows how malware visualisation can be an efficient tool for malware detection.

The domain of malware detection has to comply with several challenges, such as anti-emulation, obfuscation, Application analysis, feature extraction and preprocessing to the detection itself. A recent survey conducted by Bakour et al. [13] reviewed a significant amount of works from 2009 to 2019, where they developed a taxonomy addressing the most challenging problems in 200 proposed frameworks. The taxonomy highlighted certain aspects that are critical to this study's objectives. One of which concerns the type of features. The majority of the research was focused on Static features for mobile malware detection, which is why this study seeks to investigate whether these considerations will impact the efficiency of the proposed models. Additionally, the taxonomy investigated the techniques used for the proposed solutions. And we could see that the majority of the work is based on supervised Machine Learning.

It should be noted that much of the data generated in the real environment is either unlabelled or semi-labelled, so the exploration of supervised and semi-supervised learning can be of real benefit to the field of anomaly detection. Nevertheless, at the time of writing this paper supervised models have evolved much further than the other models while consistently achieving higher results.

Summary and Research Questions

As we can see from the aforementioned works, ML is consistently providing higher levels of detection rates and most of the current approaches have reached peak performance using deep models, hence we can expect deep learning models to surpass other detection techniques. However, there is still no thorough research into other methods of machine learning algorithms, such as ensemble methods and boosted algorithms to the best of our knowledge. Furthermore, less attention has been attached to the way features are used and integrated knowing that this has a potential to improve significantly ML models (simplified models with better performance).

Therefore, we can identify a lack of research when one needs to answer questions regarding the different types of features to be used and their integration and deployment-ability. Specifically, given these gaps, the following research questions deserve to be investigated:

- Which characteristics are the most relevant markers for malware detection?
- Which machine learning models are the most effective in this detection approach?
- Which model is the best implementation for a lightweight solution?
- How robust are these models towards zero-day attacks?

This paper stresses the importance of producing quality datasets in order to benefit from these advancements in machine learning. In fact, while some proposed approaches are successful at identifying variations within their published datasets, they can become "limited" in real-world scenarios. Recently, a number of reliable datasets were developed and made publicly available. A good example is the datasets published by CIC in collaboration with UNB, where they have generated datasets that have been periodically updated over the past few years to address the major challenges discussed earlier. These datasets were used in several other projects; hence, we can consider them as a reliable benchmark for our experiments.

2.2 Data Preparation and Methodology Presentation

In order to meet the objectives outlined in this work, several datasets were considered after reviewing the most systematic and reliable set of relevant studies. The datasets published in [9] called DL-Droid [14], as "CCCS-CIC-AndMal-2020" and [2] as "CCCS-CIC-MALDROID-2020" were selected, as these datasets are the most recent and the most feature rich, enabling us to test our hypotheses using a large number of samples.

DL-DROID dataset consists of 11,505 malware samples and 19,620 Benign applications, including paid applications, utility apps, banking apps, and popular games. In addition to satisfying most of the specifications mentioned earlier, this dataset stands out as the input was generated using Droidbot as a stateful input technique, where both static and dynamic features were extracted using Dynalog. Their experiments achieved higher results using the stateful generated dataset. However, we could only get access to the stateless version of this dataset. Thus, our results will only be contrasted to those obtained using the stateless dataset. The following Fig. 1 illustrates the preparation process and evaluation method using DL-Droid for our testbed.

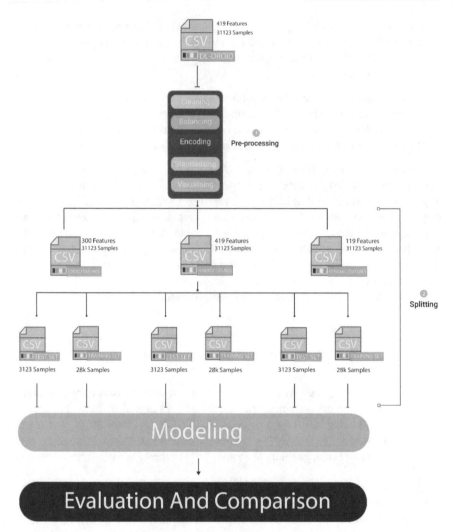

Fig. 1 DL-Droid dataset preparation and model development

Next datasets were both published by the University of New Brunswick (UNB) in collaboration with the Canadian Institute for Cybersecurity (CIC). The first Dataset "CIC MALDROID" used over 17,000 samples, where they extracted several features using Dynamic Analysis with the CopperDroid tool. The samples were collected from December 2017 to December 2018, and it includes: ·

- 17,341 Android samples of five different categories: Adware, Banking malware, SMS malware, Riskware, and Benign.
- Captured Logs: of the output analysis results of 13,077 samples for the five categories.
- Two CSV files:

  a. 470 extracted features comprising frequencies of system calls, binders, and composite behaviours.
  b. 139 extracted features comprising frequencies of system calls. Figures 2 and 3 below showcase the preparation process.

Fig. 2 CICMALDROID preparation

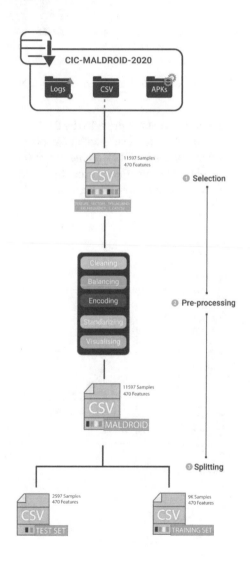

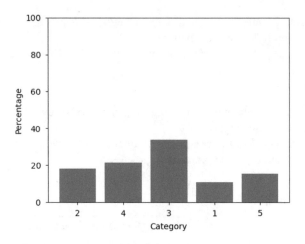

Fig. 3 Categorical distribution for the modified Maldroid dataset

The second dataset provided by the CCCS-CIC collaboration project' CCCS-CIC And Mal-2020, has a substantial sample size of 400,000 samples spanning over 14 malware classes and 191 eminent malware families including adware, backdoor, file infector, no category, potentially unwanted apps (PUA), ransomware, riskware, scareware, trojan, trojan, trojan. The next F illustrate the preparation process (Figs. 4 and 5).

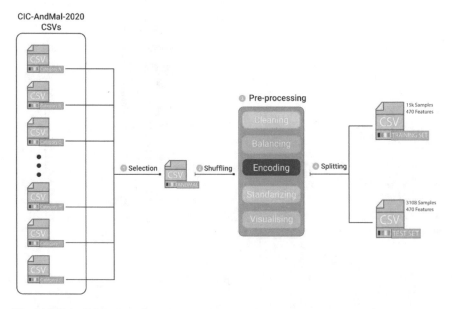

Fig. 4 CIC-AndMal preparation

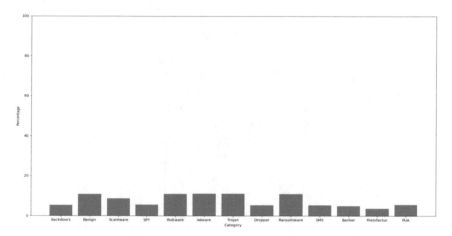

Fig. 5 Categorical distribution of the modified CIC-AndMal dataset

As mentioned, these datasets were selected since they were the closest to meet the qualities highlighted earlier; however, this selection was mainly complimentary, as some datasets cover some qualities that the others lack. For instance, the CSV files included in the CCCS-CIC datsets, only contain one type of features (Dynamic for Maldroid and only Static for AndMal), while DL-DROID contains a hybrid set of features. In the other hand, the CCCS-CIC datasets contain several types of malware categories labelled accordingly, while DL-Droid only focuses on a binary classification (Benign or Malware).

Additionally, as we can see from the preparation process, two datasets are prepared each time. A training set and a testing set. However, in deep learning models the training uses three types of datasets: a training set, a validation set, and a test set. Learning will be carried out on the training set and adjusted according to the results obtained on the validation set (that is usually 20% of the Training set). Then, the final results will be assessed by running the trained model on the test set. The following Fig. 6 illustrates the process.

This process is crucial to avoid any information leakage between the training set and the test set. As that can lead to bias outcomes that may impact the validity of the collected results.

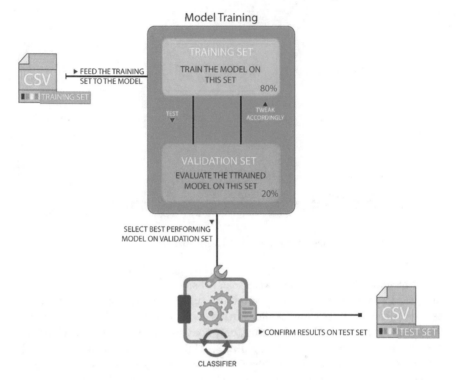

Fig. 6 Validation process

3 Proposed Models for Malware Detection

3.1 Modelling and Experimentation Plan

Figure 7 outlines the modelling and experimentation plan for our testbed.

First, an evaluation of which type of features are better indicators of the observed samples will be conducted. This would be tested using the different datasets, each with a type of feature (Static, Dynamic or Hybrid). Further, the experiments will study the benefits of feature reduction using Autoencoders, which has its own benefits of reducing feature dependencies and dimensionality reduction. The results should provide an indication of whether these techniques of dimensionality reduction will have an effect on the performance of the model.

In order to conduct this first experiment, static and dynamic features will be extracted from the DL-Droid Dataset and tested separately as seen in Fig. 1, and then later compared to the results obtained from the Hybrid dataset. Additionally, the prepared CICAndMal dataset will be used to evaluate the performance using only Static features, and later Dynamic features using CICMALDROID dataset. This would allow us contrast the results obtained from a binary classification (using

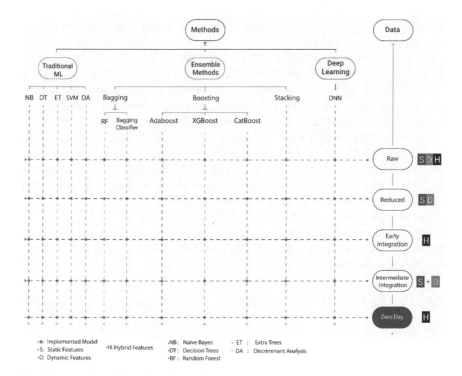

Fig. 7 Proposed models and experimental testbed

DL-Droid's dataset) to the results obtained in a multiclass classification task (using CCIS-CIC Datasets) but this time we will see how these features perform in task.

Also, to further expand on the point of "multiclass versus binary classification" performance, extra tests were carried out where the same dataset containing several malware families was experimented on; first using the different family labels in a multiclass classification, and then later, only using "malware" and "benign" labels in a binary classification.

Secondly, after reviewing the most recent papers, a set of the most popular, recent, and best performant algorithms were selected, including:

- Machine learning Traditional Algorithms: Decision Trees, Extra Trees, SVM, Discriminant Analysis and Naive Bayes.
- Ensemble Methods, including Bagging, Random forests, boosting using: Adaboost XGBoost and CatBoost. And finally Stacking were RandomForest, ExtraTrees and the Bagging Classifier were used as meta learners, and Logistic Regression as the decision-maker.
- Deep learning: using Deep Neural Networks and Autoencoders for dimensionality reduction.

Thirdly, in order to accurately test the lightweight aspect of the different models, an entire framework should be considered from feature extraction to classification, which is out of the scope of this project. However, we will execute these models in the same setup which includes a Hexacore CPU i7 8750H and a GTX 1050ti MaxQ with 16 GB of RAM, where execution times, including the training and the testing, will be compared while running on CPU only, and memory allocation will be compared to give us an indication of resource management of each model.

Finally, for the last objective, we will create a dataset using the CICAndMal dataset, where we will train the models on a set of defined malware categories, and then later test the trained model on another dataset that contains new categories. This should give an indication of how the models behave on completely new observation and new types of samples. Figure 8 illustrates the process.

These experiments will explore the research gaps addressed in this project. The results should provide a clearer approach to creating a fully cohesive and highly performant ML-based system for Mobile malware detection and prediction that can also address zero-day exploits.

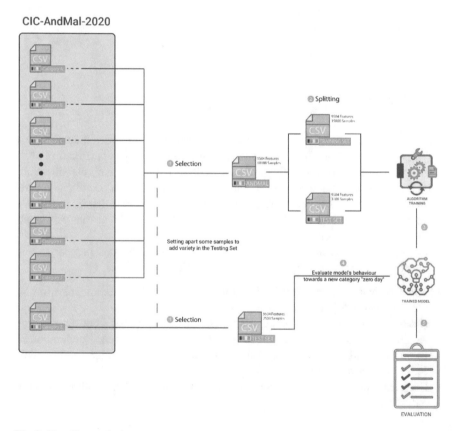

Fig. 8 Zero-Day testbed

ML Algorithms

We will first explore the traditional Machine Learning Algorithms that has proved efficient for such task. Our implementation will take advantage of the "Scikit-learn" open-source library for most models. This library offers several resources for model fitting, data pre-processing, model exploration and analysis, as it facilitates both supervised and unsupervised learning. By using the fit function, each estimator can be applied to any dataset.

- Support Vector Machine algorithms will be implemented using the "sklearn.svm" module.
- Decision trees will be implemented using the "sklearn.tree.DecisionTree Classifier" class.
- Bayes Classifier will be implanted using the "sklearn.naive_bayes" module.
- KNN Classifier will be implemented using the "sklearn.neighbours. KNeighborsClassifier" class.

Ensemble Algorithms

Contrary to ordinary approaches that train one learner, ensemble methods build on training a set of learners called Base Learners, Individual Learners or Component Learners (Depending on whether it is a homogeneous or heterogeneous set), that can be of any kind of Machine learning algorithms in order to solve a particular problem. This type of methods is interesting primarily because it is capable of improving the efficiency of the base learners [15]. Three major categories can be distinguished.

Boosting

The term Boosting refers to a family of algorithms that are able to convert weak learners into stronger learners. In our experiments, we are interested in: AdaBoost as the most influential boosting algorithm, XGBOOST as a new and popular enhanced version of the Gradient Boosting Machine, and finally CATBoost as another highly performant variation that can handle categorical data. Each algorithm was implemented using their respective libraries:

- Adaboost: using the "sklearn.ensemble.AdaBoostClassifier" implementation from Scikit-learn.
- XGBOOST: Using the XGBboost Package version 1.2.1.
- CATBoost: Using the CATBoost package version 0.24.3.

Bagging

Bagging emerges from "Bootstrap AGGregatING", as the name implies, the main components for bagging are bootstrapping and aggregation. The idea is to minimise errors by keeping the basic learners as independent as possible. This is achieved by generating basic the learners in parallel rather than sequentially. Two bagging classifiers are considered:

- The bagging classifier, available in the Scikit Library: "class sklearn.ensemble. BaggingClassifier".
- Random forest classifier, implemented using the "sklearn.ensemble. RandomForest-Classifier" class.

Stacking

Is a general procedure where a second-level Learner is trained to integrate a set of individuals (and usually heterogenous) learners known as first-level learners or metalearners. Basically, the idea is to is "combine by learning", where the original data set Is first used to train the meta learners, and then generate a new dataset from their outputs, that will be used to train the second level learner.

A Stacking algorithm can be implemented using the "sklearn.ensemble. StackingClassifier" class. Figure 9 illustrates the implemented stacked model.

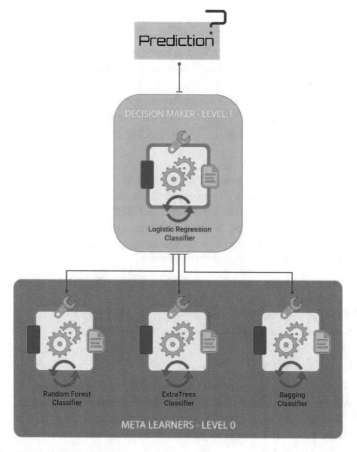

Fig. 9 Stacked model Illustration

Deep Learning Algorithms

Deep Neural Network (DNN)

Deep learning models represent the extension artificial Neural networks applied to a deep context. This extension was only made possible with the developments in calculation power, notably the GPU units. Deep learning models implement multiple layers of connected neurons fed from an input layer, relaying and processing signals to achieve a desired output. Multiple architectures can be distinguished, the most popular implementations include, recurrent neural networks (RNN), convolutional neural networks (CNN), Generative Adversarial Networks (GAN), Deep Neural Networks (DNN), Autoencoders (AE) and other some other variants of these models.

Several works have implemented Deep neural network architectures in their studies, where they could achieve higher results compared to the most established ML algorithms [9].

Deep learning problems can be solved using different architectures and different hyperparameters. Achieving the best results would require a long process of trial and error, altering the model's hyperparameters. And since this paper is restricted by a size limit, the implementation details using the TensorFlow library [16] can be found in the code repository in appendix A.

Auto-Encoders

Auto-Encoders (AEs) are artificial neural networks which take input data with a certain representation and attempts to reproduce the same data using a different representation (usually called latent representations or coding) in an unsupervised approach. Figure 10 illustrates the structure of an AE.

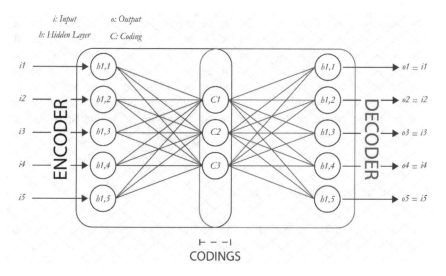

Fig. 10 Auto-encoder architecture

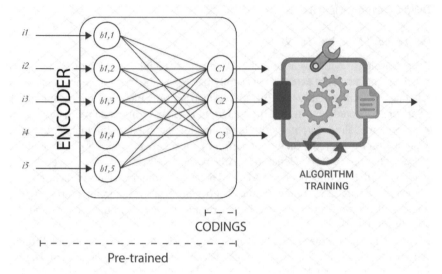

Fig. 11 Combining model with AE

In order to produce the output vector, autoencoders transform the input vector into a typically lower-dimensional representation of the data, which should provide sufficient information to generate the same input data at a fairly low error rate.

In the same fashion, these networks can also be used as feature extractors and unsupervised pre-training for other models Fig. 11.

Integration Techniques

Since most anomaly detection problems are multimodal problems, Autoencoders can be used to combine the different modalities/ types of input information in order to maximise the information gain and, consequently, the performance of the output model. By doing so, different types of integration can be used to tackle the varying noise levels and conflicts between modalities as seen in Figs. 12, 13 and 14.

These approaches would allow the incorporation of the different features obtained from both static and dynamic analysis, which theoretically should help us design a more sophisticated classifier.

4 Implementation and Evaluation

In this section, we will examine the outcomes of our experiments according to the outlined objectives. Figure 15 outlines the evaluation plan.

Fig. 12 Illustration of the early integration process

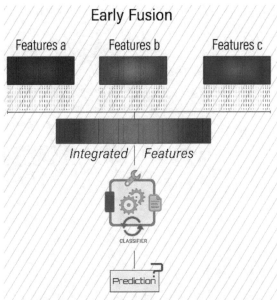

The simplest way to integrate the various feature vectors that provides a correlated representation of the different modalities.

4.1   Part 1: Features and Models

First, we start by evaluating the impact of utilising the different types of features (Static/Dynamic/Hybrid) on each model. The following tables provide the results obtained from the detection performance analysis on the DL droid dataset. Table 1 reports the results using only Static features, then Dynamic Features in Table 2, and finally, Hybrid Features in Table 3.

As seen from the tables above, most models achieved their best classification performance when both static features and dynamic features were used together in the hybrid dataset. This aligns with the given hypothesis in the problem discovery. As for the performance of each algorithm, Boosted and Tree classifiers achieved the best results in these datasets surpassing the results achieved in the original works of [9] where they achieved 0.9542 accuracy rate and 0.983771 AUC score using both features with stateless input.

On the other hand, we can see a pattern where Naïve bayes is underperforming compared to the other models, whereas in literature, it is one of the most popular algorithms. This algorithm is computationally efficient and highly capable of handling high dimensional data. However, it can perform poorly in some cases. As seen in the previous results, NB achieved the worst performance in most cases. This can be explained by the fact that the available NB classifier handles only normally distributed

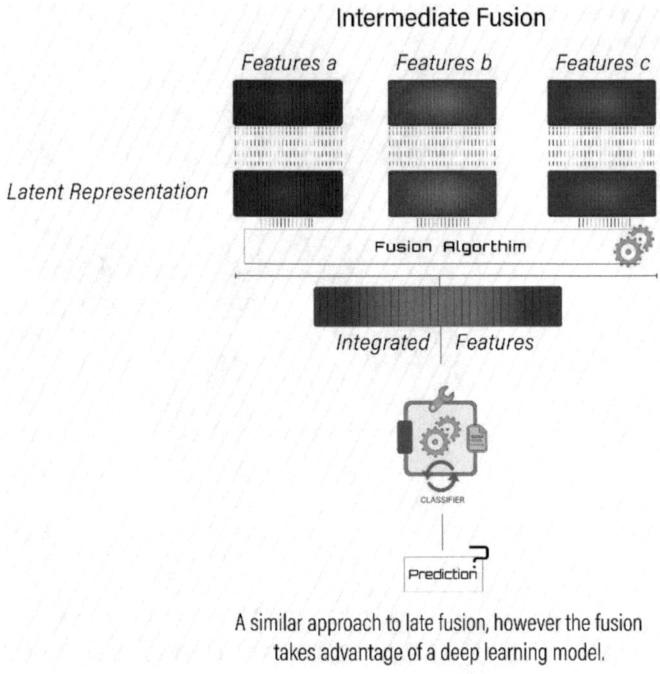

Intermediate Fusion

Features a   Features b   Features c

Latent Representation

Fusion Algorthim

Integrated | Features

CLASSIFIER

Prediction

A similar approach to late fusion, however the fusion
takes advantage of a deep learning model.

Fig. 13 Illustration of the intermediate integration process

Fig. 14 Illustration of the
late integration process

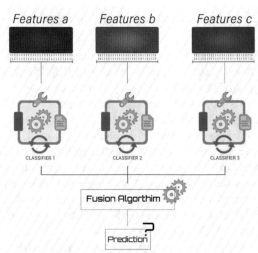

Features a   Features b   Features c

CLASSIFIER 1   CLASSIFIER 2   CLASSIFIER 3

Fusion Algorthim

Prediction

By training one model per modality and then combining
decisions with a fusion mechanism, this approach increases
flexibility and manages missing modalities (Gibert, et al,
2020).

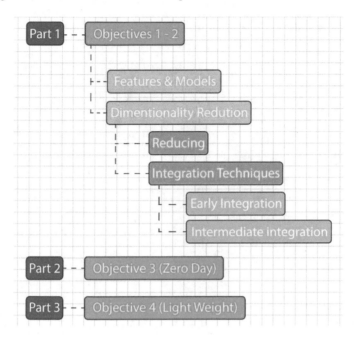

Fig. 15 Evaluation plan

Table 1 Results obtained using **Static** features from DL-Droid Dataset

Static models	Precision	Recall	f1-Score	AUC	Accuracy
Decision trees	0.9267	0.9211	0.9238	0.92114	0.9299
Random forest	0.9462	0.9364	0.9409	0.93636	0.9459
Extra trees	0.9452	0.9352	0.9399	0.93523	0.9449
Naïve Bayes	0.6812	0.5556	0.3855	0.55555	0.4361
SVM	0.9124	0.8763	0.8898	0.87625	0.902
Discriminant analysis	0.8884	0.8514	0.8648	0.85141	0.8802
Bagging	0.9513	0.9289	0.9384	0.92888	0.9443
AdaBoost	0.8747	0.8564	0.8641	0.85643	0.877
XGboost	0.9448	0.935	0.9395	0.93498	0.9446
CATboost	0.9414	0.928	0.934	0.928	0.9398
Stacking	0.9459	0.9388	0.9421	0.93881	0.9468
DNN	0.1819	0.5	0.2667	0.5	0.3638

data in case of continuous features whereas the datasets contain several features with different distributions. Furthermore, it should also be noted that NB is susceptible to correlated variables, which may have contributed to the bad performance.

Table 2 Results obtained using **Dynamic** features from DL-Droid Dataset

Dynamic models	Precision	Recall	f1-Score	AUC	Accuracy
Decision trees	0.9222	0.9129	0.9172	0.91098	0.9232
Random forest	0.9427	0.9292	0.9352	0.92925	0.9401
Extra trees	0.9454	0.9302	0.9368	0.93019	0.9417
Naïve Bayes	0.7708	0.7715	0.7712	0.77148	0.7851
SVM	0.9139	0.8831	0.8947	0.88309	0.9046
Discriminant analysis	0.8761	0.8435	0.8551	0.84348	0.8694
Bagging	0.9456	0.9326	0.9384	0.93258	0.943
AdaBoost	0.8678	0.8464	0.8547	0.84638	0.8674
XGboost	0.9451	0.9337	0.9388	0.93369	0.9433
CATboost	0.9244	0.9115	0.9172	0.9115	0.9235
Stacking	0.9431	0.9328	0.9375	0.93283	0.942
DNN	0.9331	0.9234	0.9279	0.92344	0.9331

Table 3 Results obtained using **Hybrid** features from DL-Droid Dataset

Hybrid models	Precision	Recall	f1-Score	AUC	Accuracy
Decision trees	0.9485	0.9468	0.9476	0.94681	0.9513
Random forest	0.9692	0.9626	0.9657	0.9626	0.9683
Extra trees	0.9713	0.9647	0.9678	0.96466	0.9702
Naïve Bayes	0.6862	0.5676	0.4113	0.56759	0.456
SVM	0.9496	0.9334	0.9405	0.93339	0.9456
Discriminant analysis	0.9096	0.8942	0.9009	0.89425	0.9094
Bagging	0.9696	0.9596	0.9642	0.9596	0.967
AdaBoost	0.9071	0.8998	0.9032	0.89978	0.9107
XGboost	0.9681	0.9643	0.9662	0.9643	0.9686
CATboost	0.9578	0.9554	0.9565	0.95539	0.9597
Stacking	0.9692	0.966	0.9676	0.96604	0.9699
DNN	0.9588	0.9549	0.9568	0.95492	0.96

The results we achieved have shown a distinct performance impact of the feature type on all the models. However, as seen in the Fig. 16, the obtained results were not revealing enough whether dynamic features were more indicative than static features. However, most of the models (after removing the outliers) achieved higher accuracy rates when only static features were used rather than dynamic features.

In order to further explore this behaviour, the CCCIS-CIC datasets will be analysed in a multiclass classification.

The modified AndMal dataset used in these models only includes 20 k samples (Benign and malware). This can affect the performance of deep models, especially considering the high dimensionality of the problem, however, hardware limitation

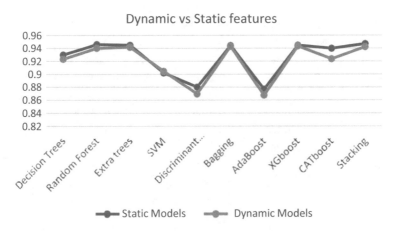

Fig. 16 Dl Droid Static versus dynamic accuracy performance

were met in certain combination. Thus, extra hyper tuning was done in this set of experiments to guarantee the achievability of best results. Table 4 reports the results achieved using the CICMALDROID Dataset that uses Dynamic features, and Table 5 reports the results on CICANDMAL that contains the Static features (Figs. 17, 18 and 19).

The implemented models were fine-tuned in both experiments (DL droid and CCIS-CIC datasets). A pattern was observed from these experiments where Static features were in most cases much more indicative than dynamic features as seen in Fig. 20, this time with an average accuracy performance increase of 12% on average.

Table 4 Results obtained using **CIC-MalDroid** Dataset

Classification models	Precision	Recall	f1-Score	AUC	Accuracy
Decision trees	0.5256	0.4722	0.4399	0.660883	0.4128
Random forest	0.666	0.6437	0.6336	0.775289	0.6257
Extra trees	0.7941	0.7986	0.7816	0.870773	0.772
Naïve Bayes	0.2861	0.221	0.0953	0.51227	0.174
SVM	0.7811	0.7438	0.7577	0.843486	0.7863
Discriminant analysis	0.6757	0.6543	0.655	0.787642	0.6916
Bagging	0.6173	0.6269	0.5691	0.761752	0.571
AdaBoost	0.6302	0.4701	0.4238	0.666953	0.4559
XGboost	0.7301	0.6278	0.6419	0.764542	0.6276
CATboost	0.7957	0.7564	0.76	0.846732	0.7528
Stacking	0.7617	0.7511	0.7246	0.840029	0.7127
DNN	0.8763	0.862	0.8684	0.91693	0.8895

Table 5 Results obtained using **CIC-AndMal** Dataset

Classification models	Precision	Recall	f1-Score	AUC	Accuracy
Decision trees	0.7873	0.7697	0.7724	0.876341	0.7973
Random forest	0.8628	0.8551	0.8561	0.921934	0.8665
Extra trees	0.8756	0.8629	0.8645	0.926192	0.8745
Naïve Bayes	0.479	0.1947	0.1487	0.565773	0.2619
SVM	0.7609	0.7283	0.7313	0.854677	0.7754
Discriminant analysis	0.7616	0.7622	0.7536	0.872044	0.7809
Bagging	0.8613	0.8505	0.8529	0.919646	0.8668
AdaBoost	0.557	0.399	0.3955	0.676223	0.4501
XGboost	0.7739	0.7322	0.7347	0.856131	0.7629
CATboost	0.6711	0.5974	0.6107	0.784892	0.6718
Stacking	0.8727	0.8577	0.8617	0.923412	0.8703
DNN	0.8452	0.836	0.8351	0.912	0.8571

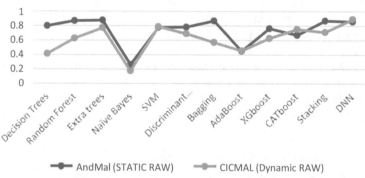

Fig. 17 CCIS-CIC accuracy performance

```
stacked_encoder = tf.keras.models.Sequential([
tf.keras.layers.Dense(input_shape=[len(featureMatrixTR[0])],units_=800, name="SAE_Layer_1"),
tf.keras.layers.Dense(units=40, activation="relu", name="SAE_Layer_2_Features"),

])
stacked_decoder = tf.keras.models.Sequential([
tf.keras.layers.Dense(units=800, activation="relu", input_shape=[40], name="SAE_Layer_3"),
tf.keras.layers.Dense(units=_len(featureMatrixTR[0]), activation="sigmoid",name="SAE_Layer_4"),
])
stacked_ae = tf.keras.models.Sequential([stacked_encoder, stacked_decoder])

stacked_ae.compile(loss="mse",
optimizer=tf.keras.optimizers.Adam(),
        )
```

Fig. 18 AndMal SAE implementation

```
stacked_encoder = tf.keras.models.Sequential([
tf.keras.layers.Dense(input_shape=[len(featureMatrixTR[0])],units_=300, name="SAE_Layer_1"),
tf.keras.layers.Dense(units=100, activation="relu", name="SAE_Layer_2_Features"),

])
stacked_decoder = tf.keras.models.Sequential([
tf.keras.layers.Dense(units=300, activation="relu", input_shape=[100], name="SAE_Layer_3"),
tf.keras.layers.Dense(units=_len(featureMatrixTR[0]), activation="sigmoid",name="SAE_Layer_4"),
])
stacked_ae = tf.keras.models.Sequential([stacked_encoder, stacked_decoder])

stacked_ae.compile(loss="mse",
optimizer=tf.keras.optimizers.Adam(),
                   )
```

Fig. 19 Maldroid SAE implementation

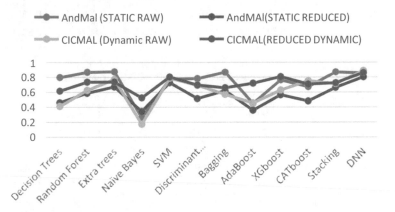

Fig. 20 Feature reduction impact on CCIS-CIC datasets

4.2 Part 1: Dimensionality Reduction

Reducing

The next set of experiments will also build on the same objective. However, this time the focus of the experiments will be on the impact of stacked autoencoders as a mean of feature reduction on the performance of each model.

The auto-encoder used in the AndMal dataset uses a two-layer encoder that reduces the number of features from 9009 to 40, Fig. 81. As for the Maldroid SAE it uses a similar architecture, though reducing the features down to 100 from 470, Fig. 19.

Tables 6 and 7 and Fig. 20 report the results achieved on AndMal and Maldroid datasets.

First, it should be mentioned that the SAE used for the AndMal models has reduced the number of features down to 40 features which represent around 0.5% of the total

Table 6 Results obtained using **reduced Features** from **CIC-AndMal** Dataset

Reduced models	Precision	Recall	f1-Score	AUC	Accuracy
Decision trees	0.4035	0.3893	0.3831	0.671958	0.4617
Random forest	0.5511	0.5072	0.4866	0.736372	0.5869
Extra trees	0.6607	0.6129	0.6059	0.792491	0.6718
Naïve Bayes	0.3959	0.3629	0.2876	0.654244	0.343
SVM	0.719	0.6893	0.6872	0.833204	0.7275
Discriminant analysis	0.5523	0.4973	0.4926	0.728298	0.5151
Bagging	0.6043	0.5697	0.5652	0.76888	0.619
AdaBoost	0.332	0.305	0.2908	0.625625	0.3581
XGboost	0.5639	0.5234	0.5159	0.743238	0.564
CATboost	0.4448	0.4392	0.4324	0.697784	0.4817
Stacking	0.6873	0.6078	0.6056	0.789741	0.6618
DNN	0.7929	0.7814	0.776	0.882429	0.8021

Table 7 Results obtained using **reduced Features** from **CIC-MalDroid** Dataset

Reduced class. models	Precision	Recall	f1-Score	AUC	Accuracy
Decision trees	0.5995	0.611	0.6003	0.75684	0.6161
Random forest	0.7294	0.7344	0.718	0.83503	0.7351
Extra trees	0.7873	0.7549	0.7446	0.84546	0.737
Naïve Bayes	0.5457	0.4657	0.4499	0.66875	0.5248
SVM	0.7938	0.7648	0.775	0.85654	0.8032
Discriminant analysis	0.6987	0.6439	0.6585	0.78095	0.695
Bagging	0.6961	0.6776	0.6585	0.79786	0.6577
AdaBoost	0.6819	0.6633	0.6692	0.7958	0.7181
XGboost	0.798	0.7896	0.7865	0.8708	0.8063
CATboost	0.7175	0.7223	0.7043	0.8244	0.7062
Stacking	0.7743	0.7472	0.731	0.84019	0.7235
DNN	0.8566	0.84	0.8468	0.90339	0.8602

recorded features in this dataset, and that is mainly due to hardware limitations. Other architectures and turn arounds were therefore tested; however, that did not effectively impact the loss rate; hence, experiments were carried out with this light SAE version. As we can see, this SAE impacted the performance of all the models negative ly, achieving worse results on the AndMal dataset compared to non-reduced models.

Nevertheless, we could see a substantial increase in accuracy rates on the MALDROID dataset when we combined the models with the respective pre-trained SAE. In fact, an increase of almost 10% of the average accuracy rate of all models was observed in this multiclass classification.

```
stacked_encoder = tf.keras.models.Sequential([
tf.keras.layers.Dense(input_shape=[len(featureMatrixTR[0])],units_=60, name="SAE_DynamicLayer_1"),
tf.keras.layers.Dense(units=30, activation="relu", name="SAE_DynamicLayer_2_Features"),

])
stacked_decoder = tf.keras.models.Sequential([
tf.keras.layers.Dense(units=60, activation="relu", input_shape=[30], name="SAE_DynamicLayer_3"),
tf.keras.layers.Dense(units=_len(featureMatrixTR[0]), activation="sigmoid",name="SAE_DynamicLayer_4"),
])
stacked_ae = tf.keras.models.Sequential([stacked_encoder, stacked_decoder])

stacked_ae.compile(loss="mse",
optimizer=tf.keras.optimizers.Adam(),
)
```

Fig. 21 DL-Droid dynamic set SAE implementation

```
stacked_encoder = tf.keras.models.Sequential([
tf.keras.layers.Dense(input_shape=[len(featureMatrixTR[0])],units_=150, name="SAE_StaticLayer_1"),
tf.keras.layers.Dense(units=70, activation="relu", name="SAE_StaticLayer_2_Features"),

])
stacked_decoder = tf.keras.models.Sequential([
tf.keras.layers.Dense(units=150, activation="relu", input_shape=[70], name="SAE_StaticLayer_3"),
tf.keras.layers.Dense(units=_len(featureMatrixTR[0]), activation="sigmoid",name="SAE_StaticLayer_4"),
])
stacked_ae = tf.keras.models.Sequential([stacked_encoder, stacked_decoder])

stacked_ae.compile(loss="mse",
optimizer=tf.keras.optimizers.SGD(lr_=0.4),
)
```

Fig. 22 DL-Droid static set SAE implementation

The same experiments were conducted on the DL droid dataset, Figs. 21 and 22 showcase the architecture used for each data type.

The implemented auto-encoders reduced the number of features from 300 to 70 in the static features, and from 119 to 30 in dynamic features. Results are reported in Tables 8 and 9.

We can see a positive impact in both cases for most models, considering that these models used datasets ¼ the size of the original datasets. In fact, this feature reduction had a positive impact on the outlying models as we can see from the Tables 8 and 9 and Figs. 23 and 24, especially for the DNN model where we have seen a big performance drop when used with raw features.

Integration Techniques

Following on the topic of feature reduction, the next set of experiments were mainly inspired from [17], where the researchers used an intermediate integration technique to combine gene expression and transcriptome alternative splicing profiles data in order to identify cancer subtypes. In their experiments, they have consistently outperformed other PCA methods in all tested datasets. Hence, we will implement both early and intermediate integration, in efforts to achieve similar outcomes.

Table 8 Results obtained using **Reduced Static** features from DL-Droid Dataset

Static models	Precision	Recall	f1-Score	AUC	Accuracy
SAE-decision trees	0.8549	0.8222	0.834	0.82223	0.8527
SAE-random forest	0.9406	0.9288	0.9341	0.92875	0.9398
SAE-extra trees	0.9412	0.9282	0.934	0.92819	0.9398
SAE-Naïve Bayes	0.683	0.6945	0.661	0.6945	0.6628
SAE-SVM	0.9027	0.8702	0.8826	0.87021	0.8953
SAE-Bagging	0.9392	0.9259	0.9319	0.92592	0.9379
SAE-AdaBoost	0.8919	0.879	0.8847	0.87901	0.895
SAE-XGboost	0.9384	0.9302	0.934	0.9302	0.9395
SAE-CAT	0.9464	0.9455	0.9459	0.94554	0.9497
SAE-stacking	0.94	0.9314	0.9354	0.93139	0.9408
SAE-DNN	0.9406	0.9294	0.9345	0.92938	0.9401

Table 9 Results obtained using **Reduced Dynamic** features from DL-Droid dataset

Dynamic models	Precision	Recall	f1-Score	AUC	Accuracy
SAE-decision trees	0.8746	0.8644	0.8689	0.86437	0.8786
SAE-random forest	0.932	0.9172	0.9236	0.91722	0.9296
SAE-extra trees	0.9344	0.9128	0.9399	0.9217	0.9283
SAE-Naïve Bayes	0.6796	0.6773	0.6784	0.67734	0.7003
SAE-SVM	0.9197	0.8957	0.9053	0.89571	0.9135
SAE-bagging	0.9187	0.9041	0.9104	0.90407	0.9174
SAE-AdaBoost	0.8735	0.8674	0.8641	0.8703	0.8793
SAE-XGboost	0.9327	0.9179	0.9395	0.9243	0.9302
SAE-CATboost	0.9191	0.9084	0.934	0.9132	0.9196
SAE-stacking	0.9305	0.9186	0.9239	0.91858	0.9296
SAE-DNN	0.9378	0.9111	0.922	0.91115	0.9299

For the early integration, we have implemented an SAE that reduces the number of hybrid features down to 100 from 419. Figure 25 shows the implementation of this SAE.

As for late integration the same autoencoders used in the static and dynamic feature reduction were combined using another autoencoder, which encodes both latent representations of each feature type into a new dataset that contains 100 features. Figure 26 showcases an example architecture that feeds the outcome of the late integration directly to a DNN model.

Results are shown in Tables 10 and 11 and Fig. 27.

Unlike the works of [17] intermediate integration did not achieve similar outcomes in our experiments. in fact, the models performed the worst using this method of integration.

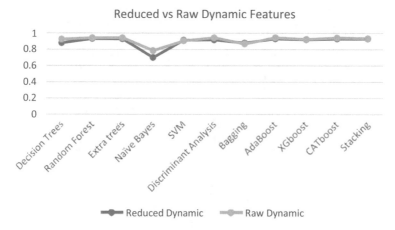

Fig. 23 Feature reduction impact on Dl-Droid's dynamic features

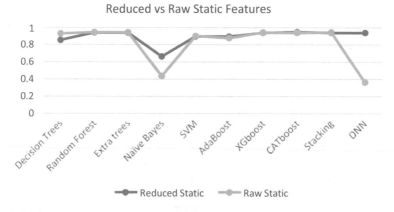

Fig. 24 Feature reduction impact on Dl-Droid's static features

```
stacked_encoder = tf.keras.models.Sequential([
tf.keras.layers.Dense(input_shape=[len(featureMatrixTR[0])],units_=200, name="SAE_HybridLayer_1"),
tf.keras.layers.Dense(units=100, activation="relu", name="SAE_HybridLayer_2_Features"),

])
stacked_decoder = tf.keras.models.Sequential([
tf.keras.layers.Dense(units=200, activation="relu", input_shape=[100], name="SAE_HybridLayer_3"),
tf.keras.layers.Dense(units_=_len(featureMatrixTR[0]), activation="sigmoid",name="SAE_HybridLayer_4"),
])
stacked_ae = tf.keras.models.Sequential([stacked_encoder, stacked_decoder])

stacked_ae.compile(loss="mse",
optimizer=tf.keras.optimizers.SGD(lr_=0.4),
                )
```

Fig. 25 Early integration SAE

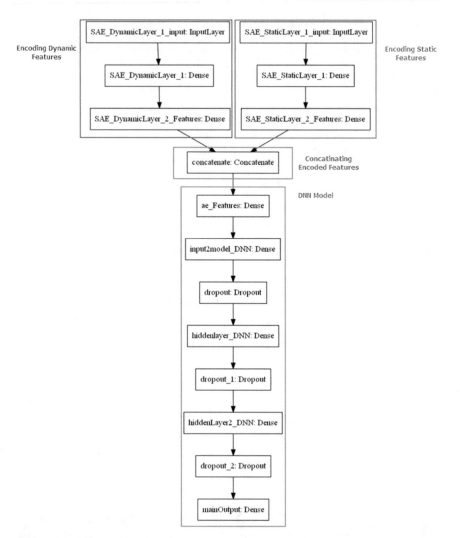

Fig. 26 Intermediately integrated DNN

However, early integration increased the average general performance across all models by more than 1% while using less than ¼ of the original hybrid dataset. In fact, we reached a peak score using this combination when integrated with the Extra Trees Classifier. In which we improved the results obtained in the original works of [9] using the stateless dataset, and then our competitive results using the raw data, but also exceeded their best scores achieved on the stateful dataset (0.985 accuracy rate and 0.997 AUC score) as we achieved 0.99913 on the AUC score and 0.9994 accuracy rate.

Table 10 Results obtained from **Early fusion** on the **DL-Droid** dataset

Hybrid models	Precision	Recall	f1-Score	AUC	Accuracy
Decision trees	0.9024	0.9013	0.9018	0.90133	0.9087
Random forest	0.9598	0.9512	0.9552	0.9512	0.9587
Extra trees	0.9995	0.9991	0.9991	0.99913	0.9994
Naïve Bayes	0.6869	0.7008	0.6811	0.70077	0.6865
SVM	0.9402	0.9208	0.929	0.92075	0.9353
Discriminant analysis	0.8901	0.8759	0.882	0.87585	0.8921
Bagging	0.9562	0.9439	0.9495	0.94388	0.9536
AdaBoost	0.9008	0.8976	0.8992	0.89757	0.9065
XGboost	0.9612	0.954	0.9574	0.95398	0.9606
CATboost	0.9426	0.9425	0.9425	0.94247	0.9465
Stacking	0.9601	0.9551	0.9575	0.95507	0.9606
DNN	0.9548	0.9536	0.9542	0.95362	0.9574

Table 11 Results Obtained from **Late Fusion** on the **DL-Droid** Dataset

Hybrid models	Precision	Recall	f1-Score	AUC	Accuracy
Decision trees	0.5023	0.5022	0.5022	0.50225	0.537
Random forest	0.4793	0.4952	0.4211	0.4952	0.6119
Extra trees	0.4777	0.496	0.4134	0.49604	0.6164
Naïve Bayes	0.4655	0.4991	0.3909	0.49906	0.6289
SVM	0.4406	0.4993	0.3882	0.49935	0.6302
Discriminant analysis	0.3157	0.5	0.387	0.5	0.6314
Bagging	0.5064	0.5028	0.4585	0.50276	0.6045
AdaBoost	0.5326	0.5035	0.4125	0.50354	0.6279
XGboost	0.4928	0.496	0.465	0.49601	0.5876
CATboost	0.5056	0.5052	0.5042	0.50523	0.5517
Stacking	0.3157	0.5	0.387	0.5	0.6314
DNN	0.3157	0.5	0.387	0.5	0.6314

Furthermore, we can see that in all feature types and datasets, early integration has decreased if not eliminated the performance drop of Naïve Bayes, which can confirm the assumptions raised earlier regarding the dependency of features, as autoencoders can handle this issue by generating new representative features.

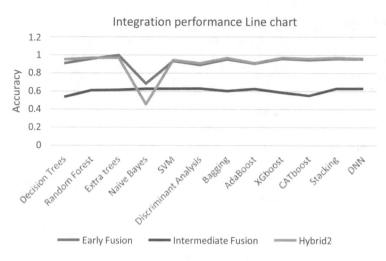

Fig. 27 Integration performance comparison

4.3 Part 2: Zero-Day Detection Efficiency

In the second part of the testbed, we tested the robustness of the chosen models according to the plan outlined in Figs. 7 and 8. Two experiments were conducted in each test; the first one using raw features, and the second using a reduced set of features. In these experiments, the dataset was modified were the zero-day samples were labelled "Malware" same as all the other Malware categories to measure the performance in a binary classification. Results are reported in Tables 12 and 13 and Fig. 28.

Table 12 Zero-day classification results using **Raw features** from **CIC-AndMal** dataset

Raw models	Precision	Recall	f1-Score	AUC	Accuracy
Decision trees	0.9957	0.9905	0.9931	0.9905	0.9956
Random forest	0.9576	0.9657	0.9616	0.96575	0.9752
Extra trees	0.9881	0.9558	0.9709	0.95575	0.982
Naïve Bayes	0.8593	0.5737	0.5809	0.57375	0.8268
SVM	0.9855	0.9447	0.9635	0.94475	0.9776
Discriminant analysis	0.9729	0.9883	0.9803	0.98825	0.9872
Bagging	0.9955	0.9932	0.9944	0.99325	0.9964
AdaBoost	0.9947	0.9865	0.9905	0.9865	0.994
XGboost	0.9982	0.9968	0.9975	0.99675	0.9984
CATboost	0.9967	0.9908	0.9937	0.99075	0.996
Stacking	0.9952	0.9923	0.9937	0.99225	0.996
DNN	0.9827	0.9545	0.9678	0.9545	0.98

Table 13 Zero-day classification results using **Reduced features** from **CIC-AndMal** dataset

Reduced models	Precision	Recall	f1-Score	AUC	Accuracy
Decision trees	0.8378	0.71	0.7478	0.71	0.8672
Random forest	0.8769	0.7423	0.7845	0.74225	0.8852
Extra trees	0.9183	0.8485	0.8776	0.8485	0.928
Naïve Bayes	0.6738	0.577	0.5874	0.577	0.8032
SVM	0.9326	0.9025	0.9165	0.9025	0.9484
Discriminant analysis	0.8744	0.6295	0.6618	0.6295	0.8476
Bagging	0.8509	0.7195	0.759	0.7195	0.8728
AdaBoost	0.8882	0.797	0.8317	0.797	0.9044
XGboost	0.8458	0.6847	0.7244	0.68475	0.8616
CATboost	0.9073	0.8203	0.8545	0.82025	0.9164
Stacking	0.9018	0.7885	0.829	0.7885	0.9052
SAE-DNN	0.9282	0.8922	0.9087	0.89225	0.944

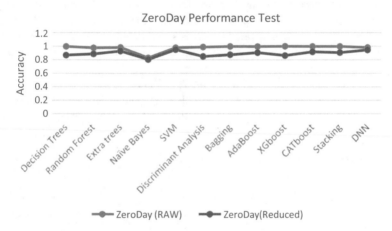

Fig. 28 Impact of feature reduction on zero-day classification

Most of the models have unexpectedly achieved high accuracy rates, wherein some instances (Boosted Models) we achieved more than 99% accuracy using Raw features with a very low false-positive rate, as can be seen from AUC scores. However, in contrast to the results obtained from the previous experiments, dimensionality reduction has decreased the accuracy rate of all the models when exposed to a new set of malware categories. Hence, a trade-off between higher accuracy rates and reliability must be measured to accommodate the issue of zero-day exploits.

4.4   Part 3: Lightweight Assessment

As a reminder, this assessment's objective is to benchmark the algorithms themselves, not a full fledged detection solution. Hence only certain characteristics of resource management were selected. However, for the sake of validity, this experiment only evaluated the most accurate characteristics that could be reliably measured. For instance, live memory usage in all, training testing and idle modes was intended to be examined. However, this characteristic can be heavily impacted by several other environmental factors, such as: the configuration platform, background services, pagination and resource sharing techniques, etc., which were entirely out of our control. So, in efforts to make the measurements as accurate as possible, a profiler was implemented to execute the tests several times in order to achieve better precision, while eliminating any influence of other background tasks such as Disk Flushing, OS scheduling and garbage collecting, that can run at any inopportune moment.

Since the best results were achieved on the Hybrid DL droid dataset, the corresponding models were selected for this experimental setup. Training and testing times were measured on the same datasets (28,000 and 3123 samples for training and testing respectively) in each model, including the SAE learning and encoding performance. Finally, the generated model is saved in either HDF5 (.h5) format for deep models, or using pickle as ".sav" format, where the memory allocation was measured afterwards.

Results are shown in Table 14 and Figs. 29, 30 and 31.

From the above statistics, we can see a clear advantage for tree classifiers when it comes to training times. As we can see, these models (apart from SVM) can be

Table 14 Resource management comparison results

Hybrid models	Time		Iterations	Allocation	
	Training (min)	Testing (s)		Size (Mbytes)	Format
Decision trees	12.487	0.0449	10	0.246794	.sav
Random forest	1.224	0.828	10	40.937841	.sav
Extra trees	5.0392	2.986	10	163.909895	.sav
Naïve Bayes	0.0456	0.3959	10	0.014091	.sav
SVM	26.3669	102.037	10	22.121703	.sav
Discriminant analysis	0.4419	0.1763	10	0.01767	.sav
Bagging	396.504	6.0936	10	25.347164	.sav
AdaBoost	4.5049	8.379	10	0.100306	.sav
XGboost	5.083	0.1017	10	2.024482	.sav
CATboost	0.9863	0.0872	10	0.44666	.sav
Stacking	30.3699	3.4502	10	174.196625	.sav
DNN	27.8126	0.923	10	0.787464	.h5
SAE (Hybrid)	28.279	1.1257	10	0.852324	.h5

Fig. 29   Training time comparison results

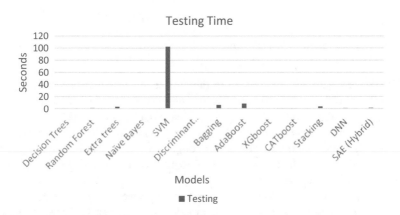

Fig. 30   Testing time comparison results

trained at about 5% (on average) of the time relative to deep learning models, while at the same time achieving competitive accuracy rates (Again, this difference can be affected by hyperparameters tuning in both respects).

Since boosted algorithms use multiple "week learners" we could anticipate the training time results, where Boosted models took more than 250% (on average) of the training time of ML models and only 15% of the DNN models' time.

The actual trade-off appeared on allocation size, where we can see that boosted algorithms maintained their balance of higher results while allocating limited resources.

A different pattern can be observed on testing durations, however this time with a narrower margin, where we can see that the boosted algorithms have taken the most time predicting the same results.

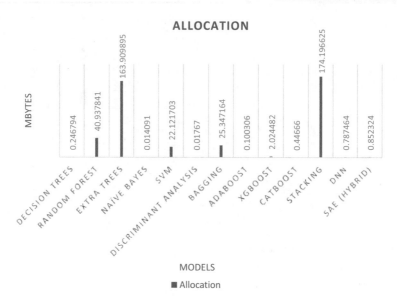

Fig. 31 Memory allocation comparison results

We can see that ML models were the fastest when it comes to training and testing times, though, when it comes to resource allocation, these models were the most resource demanding compared to all other models. In fact, the highest scoring algorithm "Extra Trees classifier" allocates more than 20,800% memory space than our proposed DNN model that achieved competitive results. In the other hand the boosted models (excluding Stacked model) exhibited the best compromise between performance, time, and allocation Size.

The proposed stacking model has consistently achieved one of the highest scores in almost all experiments. However, this model is, as we can see from the figures above, the most complex, and the most resource-demanding algorithm.

Further, the proposed feature reduction model that uses an early integration technique Adds an extra layer of complexity in these experiments. Especially when the features reduction model by itself, can in some instances allocate more space, and can take longer to train than the predictive model itself.

These findings advocate Boosted models as the right combination of detection rates and resource allocation for a lightweight and robust malware detection solution. That being said, further tests are required in the deployment environment, including all components of the detection system in order to decide whether an end-device or a server solution is superior for real-time detection.

4.5 Summary of Results

So, as a recap for these experiments, the following summarises the most notable findings according to our objectives (Figs. 32 and 33).

- **OBJECTIVE 1 (Feature Types):**

  Best Results were achieved when both static and dynamic features were combined to detect the anomalies in both binary and multiclass classification.

  Features types variation has a more significant impact on multiclass classification. Static features were more indicative than dynamic features.

- **OBJECTIVE 2 (Models):**

  ML and boosted models consistently performed well over different datasets; however, the performance of Deep Neural Networks was enhanced as soon as we loaded more data. It should be mentioned that there is still room for improvement for deep models with further hyper-parameter tuning.

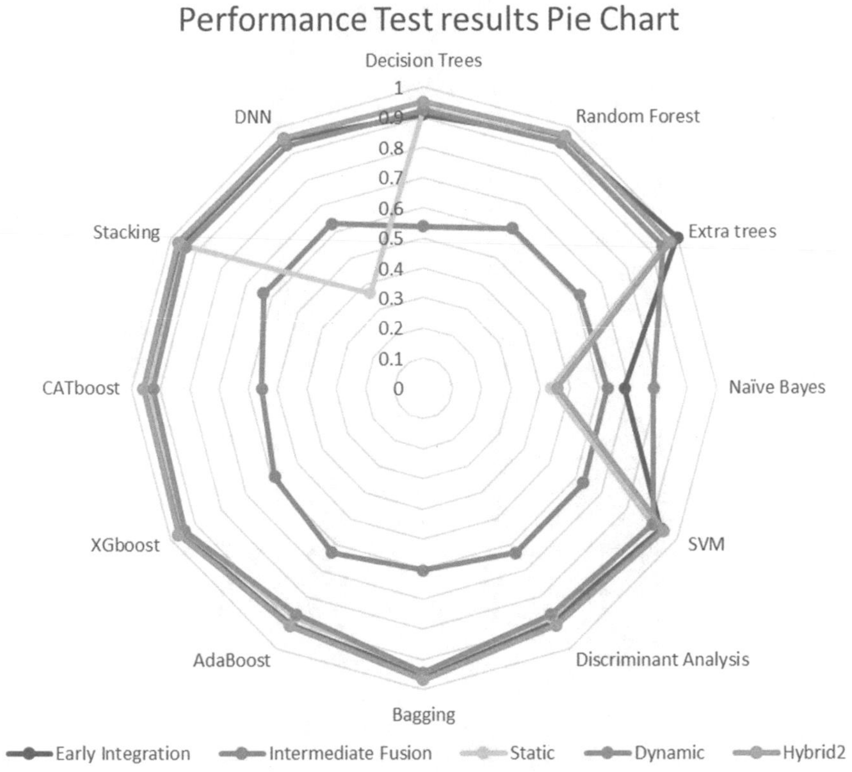

Fig. 32 Accuracy Pie chart comparison of the different techniques using DL-Droid dataset

Models	Binary Classification				Multiclass Classification				Zero Day	Lightweight		
	DL_Droid Hybrid RAW	DLDROID Early Integration	DLDROID Late Integration	Average	AndMal (Static)	MalDroid (Dynamic)	MALDROID (Reduced)	Average	RAW Data	Size Allocation (Mbytes)	Training (Seconds)	Testing (Seconds)
Decision Trees	0.9513	0.9087	0.537	0.799	0.7973	0.4128	0.6161	0.60873	0.9956	0.247	12.487	0.045
Random Forest	0.9683	0.9587	0.6119	0.8463	0.8665	0.6257	0.7351	0.74243	0.9752	40.938	73.413	0.828
Extra trees	0.9702	0.9994	0.6164	0.862	0.8745	0.772	0.737	0.7945	0.982	163.910	302.353	2.987
Naïve Bayes	0.455	0.6865	0.6289	0.590047	0.2619	0.174	0.5248	0.320023	0.8268	0.014	2.737	0.396
SVM	0.9456	0.9353	0.6302	0.83703	0.7754	0.7863	0.8032	0.7883	0.9776	22.122	1582.015	102.037
Discriminant Analysis	0.9094	0.8921	0.6314	0.81097	0.7809	0.6916	0.695	0.7225	0.9872	0.018	26.515	0.176
Bagging	0.967	0.9536	0.6045	0.8417	0.8668	0.571	0.6577	0.6985	0.9964	25.347	396.504	6.094
AdaBoost	0.9107	0.9065	0.6279	0.81503	0.4501	0.4559	0.7181	0.54137	0.994	0.100	270.292	8.379
XGboost	0.9686	0.9606	0.5876	0.83893	0.7629	0.6276	0.8063	0.73227	0.9984	2.024	304.982	0.102
CATboost	0.9597	0.9465	0.5517	0.8193	0.6718	0.7528	0.7062	0.71027	0.996	0.447	59.178	0.087
Stacking	0.9699	0.9606	0.6314	0.85397	0.8703	0.7127	0.7235	0.76883	0.996	174.197	1822.192	3.450
DNN	0.96	0.9574	0.6314	0.8496	0.8571	0.8895	0.8602	0.86893	0.98	0.787	1668.756	0.923

Fig. 33 Performance (Accuracy) summary table

We have achieved better scores compared to existing methods in literature so far with the DL-Droid dataset using Extra Trees classifier in combination with a stacked autoencoder in an early integration (99.94% accuracy).

Auto-encoders are an efficient tool for dimensionality reduction that can sustain high performing models, if not increasing the efficiency in some cases -especially with models that can get impaired by feature dependency.

- **OBJECTIVE 3 (Zero-Day Detection Efficiency):**

High and reliable detection rates of zero-day vulnerabilities can be achieved using various models of Machine Learning Paradigms with adequate tuning. However, a compromise in accuracy rates must be regarded if feature reduction is an essential component of the proposed model.

- **OBJECTIVE 4 (Lightweight Assessment):**

Boosted models provide the best balance of detection rates and resource allocation for a lightweight and robust malware detection solution.

With an adequate amount of data and hyperparameter tuning, deep models can be superior to other conventional models in both detection rates and complexity.

Feature reduction using SAEs is a heavy pre-processing phase. Thus, an analysis of requirements should be carried out before considering this approach for complexity optimisation.

5 Conclusion and Future Work

In this work, mobile malware detection problem was tackled using traditional and advanced machine learning models on the three recent datasets: DL-Droid, CCIS-CIC-MalDroid-2020, CCIS-CIC-AndMal-2020.

Firstly, the background and an in depth review of recent and effective detection approaches were explored.

Secondly, a plan was put in place for modeling and experimentation according to which various models were designed, implemented, and evaluated. Each proposed model can be viewed as a combination of the following criteria: type of features (static vs dynamic vs Hybrid features, Raw vs Encoded), the integration strategy (early vs late feature integration) and finally the classifier. We were able to see from the obtained results that the type of classification may have a significant impact on the performance of the proposed models. However, the main findings related to our four objectives can be summarised as follows:

Objective 1: Hybrid features derived from static and dynamic analysis made it pos sible for the models to achieve their highest accuracy rates compared to other types of features.

Objective 2: The implemented models behaved differently towards the different types of features and integration techniques. Many models have outperformed the currently proposed approaches in literature, however a particular combination of early integration with the Extra trees Classifier achieved the highest accuracy rate (of 99.94%) in this testbed.

Objective 3: Boosted algorithms offered the best balance of detection rates and resource utilisation for a lightweight and robust malware detection solution.

Objective 4: The proposed ML-Models have in most cases showcased a robust detection efficiency against new malware categories, achieving more than 99% accuracy rate in certain instances. In the other hand, we also observed a negative impact of feature reduction on their efficiency towards these samples.

In the future, this research can be extended in many directions:

- First, to fully take advantage of the current developments in the Machine Learning field, there will be a need to develop improved datasets to address new malware threats. They fulfil the requirements set out in the previous discussion (Reference the previous paper).
- Future studies may also further explore the optimisation of these models without impacting their performance. As we have shown from the results, dimensionality reduction increased the accuracy of several models by using a fraction of the dataset, but it has limited the accuracy of zero-day attack prediction. Therefore, we can see that there are opportunities for advancement in this area where we could develop a more optimised and robust solution.
- Recent advances in malware have resulted in enhanced obfuscation and evasion mechanisms, thus, studying how the various models respond to this behaviour can be very insightful, especially since most of the available data is derived from static analysis.
- Semi-supervised and unsupervised models can be intriguing to explore, provided that most of the real-world machine learning problems fall into this category and most of the available data is either labelled or semi-labelled.
- Further experiment on the lightweight aspect of these models could be conducted, where live memory usage and power consumption is measured in an appropriate environment to accurately check whether the new devices are capable of hosting such detection systems. Especially given the need to develop a solution that is capable of real-time monitoring and evaluation.
- One lingering problem with these models is the interpretability of their actions. Much of the technologies currently implemented are more interpretable and intuitive, such as rule-based and signature-based frameworks, and are more regarded by cybersecurity Analysts for this aspect [18]. Many of the models used at present are presented as black boxes which can be problematic, especially if it would not be possible to interpret these decisions in a strategic manner. Thus, overcoming this challenge can totally transform the industry.
- Finally, Attack prediction should be considered as a proactive approach and further examined and developed with the current machine learning-based approaches.

References

1. Melissa C, Ryan R (2020) IDC smartphone market share OS. IDC, The premier global market intelligence company. https://www.idc.com/promo/smartphone-market-share/os. Accessed 16 Mar 2021
2. Samani R (2020) McAfee mobile threat report. [ebook] McAfee. https://www.mcafee.com/con tent/dam/consumer/en-us/docs/2020-Mobile-ThreatReport.pdf. Accessed 16 Mar 2021
3. Liu H, Lang B (2019) Machine learning and deep learning methods for intrusion detection systems: a survey. Appl Sci 9(20):4396. https://www.mdpi.com/2076-3417/9/20/4396
4. Vinayakumar R, Alazab M, Soman KP, Poornachandran P, Venkatraman S (2019) Robust intelligent malware detection using deep learning. IEEE Access 7:46717–46738. https://doi. org/10.1109/ACCESS.2019.2906934
5. Mas'ud MZ, Sahib S, Abdollah MF, Selamat SR, Yusof R (2014) Analysis of features selection and machine learning classifier in android Malware detection. In: 2014 international conference on information science & applications (ICISA), Seoul, Korea (South), pp 1–5. https://doi.org/ 10.1109/ICISA.2014.6847364
6. Shhadat I, Hayajneh A, Al-Sharif ZA (2020) The use of machine learning techniques to advance the detection and classification of unknown Malware, Procedia Comput Sci 170:917–922. ISSN 1877-0509. https://doi.org/10.1016/j.procs.2020.03.110
7. Arshad S, Shah MA, Wahid A, Mehmood A, Song H, Yu H (2018) SAMADroid: a novel 3-level hybrid malware detection model for android operating system. IEEE Access 6:4321–4339. https://doi.org/10.1109/ACCESS.2018.2792941
8. Yoo S, Kim S, Kim S, ByunghoonKang B (2020) AI-HydRa: Advanced hybridapproach using random forest and deep learning for malware classification, vol 546, pp 420–435. Else-vier https://www.sciencedirect.com/science/article/pii/S0020025520308525. Accessed 22 Dec 2020
9. Alzaylaee M, Yerima S, Sezer S (2020) DL-Droid: deep learning based android malware detection using real devices. Comput Secur (Elsevier) 89:101663. https://www.sciencedirect. com/science/article/pii/S0167404819300161#bib0031. Accessed 3 May 2020
10. Zhang N, Tan YA, Yang C, Li Y (2021) Deep learning feature exploration for android malware detection. Appl Soft Comput 102:107069. ISSN 1568-4946. https://doi.org/10.1016/j.asoc. 2020.107069
11. Pei X, Yu L, Tian S (2020) AMalNet: a deep learning framework based on graph convolutional networks for malware detection. Comput Secur 93:101792. ISSN 0167-4048. https://doi.org/ 10.1016/j.cose.2020.101792
12. Naeem H, Ullah F, Naeem MR, Khalid S, Vasan D, Jabbar S, Saeed S (2020) Malware detection in industrial internet of things based on hybrid image visualization and deep learning model, Ad Hoc Netw 105:102154. ISSN 1570-8705. https://doi.org/10.1016/j.adhoc.2020.102154
13. Bakour K, Ünver H, Ghanem R (2019) The Android malware detection systems between hope and reality. SN Appl Sci 1(9)
14. Rahali A, Lashkari AH, Kaur G, Taheri L, Gagnon F, Massicotte F (2020) DIDroid: android Malware classification and characterization using deep image learning. In: 10th international conference on communication and network security, Tokyo, Japan
15. Zhou Z, Graepel T, Herbrich R (2012) Ensemble methods foundations and algorithms.1st edn. Cambridge, UK, Taylor & Francis Group
16. TensorFlow (2020) TensorFlow. [https://www.tensorflow.org. Accessed 25 Oct 2020
17. Guo Y, Shang X, Li Z (2019) Identification of cancer subtypes by integrating multiple types of transcriptomics data with deep learning in breast cancer. Neurocomputing324:20–30. https://www.sciencedirect.com/science/article/abs/pii/S0925231218306222?via%3Dihub. Accessed 22 Dec 2020
18. Gibert D, Mateu C, Planes J (2020) The rise of machine learning for detection and classification of malware: research developments, trends and challenges. J Netw Comput Appl 153:102526

An Approach of Applying, Adapting Machine Learning into the IDS and IPS Component to Improve Its Effectiveness and Its Efficiency

Lucky Singh and Hamid Jahankhani

Abstract The traditional intrusion detection and Intrusion prevention systems are known as "signature based", which means that they function in a similar method to a virus scanner by identifying the similar signatures for each intrusion event it detects. This specific method is very effective if the attacks are known, but for zero day attack it will not be able to identify the incoming threat. cited from (Meryem and Ouahidi in Netw Secur 8–19, 2020 [9]) IDS signature library requires constant update as the current IDS and IPS are only as good as their signatures and if there is a zero-day attack then the IDS will not be able to detect it. In cited from (Porter An attack with a previously unseen volume of 2.3). Different ML algorithms such as "Naïve Bayes, decision tree, K-Nearest Neighbors and logistic regression" are discussed and compared throughout the project. An implementation and development detection module a new method of detection module has been created. The research also consists of critical feature selection. The algorithm detection module was trained and tested with test data obtained from the clients' network. The algorithm was then implemented within the client live network and the results were retrieved again. The detection modules were trained on different ML algorithms and accuracy was then compared from all of them. From which "decision tree" is able to provide the highest accuracy result with a very misclassification rate. Till now 99% accuracy has been reached. On the other hand, if the large corpus of training data is used then the detection accuracy can increase. The primary objective was to deliver an upgraded version of IPS/IDS to clients which would be "Anomaly based" that would employ Machine learning approach to protect the client devices and network from different cyber-attacks such as zero-day attacks.

Keywords ML algorithm · Machine learning algorithm · "decision tree"

L. Singh · H. Jahankhani (✉)
Northumbria University, London, UK
e-mail: Hamid.jahankhani@northumbria.ac.uk

© The Author(s), under exclusive license to Springer Nature Switzerland AG 2021
R. Montasari and H. Jahankhani (eds.), *Artificial Intelligence in Cyber Security: Impact and Implications*, Advanced Sciences and Technologies for Security Applications,
https://doi.org/10.1007/978-3-030-88040-8_2

1   Introduction

Network attacks are increasing at an alarming rate as new technologies emerge. A high number of attacks incidents have been addressed by organizations every year. "DDS (Distributed denial of service") attacks increased as 2018 study link 11 identified there were 102 attacks between June 2018 and April 2018 specifically toward banks [1].

The attackers are using different intrusion methods such as "probing, flooding, ransomware/malware" attacks and so on. Which is used to steal sensitive and confidential information from organizations. During Covid-19 pandemic, there has been a high increase in cyber-attacks. As the author [2] describes "there has been a rise in the number of public high-profile cybersecurity incidents, the majority being ransomware attacks involving exfiltrated data being leaked".

There are different classification of intrusion detection and prevention given by researchers such as "DoS Denial of Service" attacks "(Bandwidth and Resource depletion)", "Scanning attacks (Probe), Remote to local (RCL) attacks and User to Root (U2R) intrusions". These are developed on "KDD99" dataset. The newer data sets "UNSW-NB" categories attacks in nine subsets: ", Analysis, Reconnaissance, Shellcode, worm, DoS and Exploit" [3]. These have been reviewed in detail below.

There are three types of IDS/IPS "(Intrusion Detection System)" or three methodologies part of a single component device.

"Network Intrusion detection system"

"Network Node Intrusion detection system"

"Host Intrusion Detection System" [4].

"Network-based Intrusion Detection System (NIDS) and Network Node Intrusion detection system" examines network traffic and determines normal or suspicious activity [3].

"Network Intrusion Detection System (NIDS)" these systems are implemented strategically throughout the networks, designed secure points, where the network is most vulnerable. Mainly it's engrafted over complete "subnets" and it would cross-reference any traffic transmitting with the library of "known attack signatures" These are easily implemented and difficult for attackers to distinguish for intrusions [4].

A "Network Node Intrusion detection system (NIDS) is similar to "NIDS", but with an indistinct variation that is only to "applied to one host at a time rather than the entire subnet" [4].

"A host-based Intrusion detection system" surveys independent "host/device" and alerts the network administrator, if "suspicious activities" for example editing or removing system files, executing suspicious numbers of system commands, suspicious "configuration" edits are identified [3].

These systems function in signature-based and anomaly-based methodologies.

"Signature matching based" contemplates the "incoming packets" against predetermined "patterns/signatures" and if any of the packet headers match the "patterns/signature". That "data packet" flagged as divergent. "State transition based" evaluation, controls a "state transition model of the system for the known suspicious patterns. Different branches of the model lead to a final compromised state of the machine" [5].

There are many disadvantages with the "signature-based IDS/IPS" is that a consistent stream of signatures/rules are required to detect and prevent unknown and malicious intrusions. To which it is not able to detect unknown or zero-day attacks.

"Anomaly-based detection IDS/IPS" are founded upon user behavioral patterns to differ between normal users to suspicious behaviors. "Anomaly-based IDS/IPS model function by generating a database with regular behaviors of the network and constantly updates over time. Such as "network connection and transmission" in certain features "protocols, service, the number of login attempts, packets per flow, bytes per flow, source address, destination ports, etc.". Any alterations in the feature values record of the network transmissions will be flagged up by the "anomaly detection model".

There are three different types of anomaly-based detection:

"Machine learning-based techniques: Finite state machine-based techniques:" [5].

"Finite state machine (FSM)" generates a "behavioral model" which is a collection of "states, transitions and events".

"Machine learning-based IDS/IPS" provides a learning module for the system to learn pre-existing data, discover patterns and make decisions to protect the network without the assistance from the administrator. This is highly beneficial in detecting and preventing zero-day attacks.

In this chapter, we have primarily concentrated on the use of machine learning for anomaly detection methods with detailed evaluation and examination of its attack detection capability. Having a detailed study of numerous machines learning approaches is beneficial while creating your machine learning algorithm. Machine learning intrusion detection has been divided into four categories.

"Single classifiers with all features of the data set:

Single classifiers with limited features of data set

Multiple classifiers with all features of data set

Multiple classifiers with limited features of data set"

Single classifier system: an independent "classifier" is utilized to identify intrusions. "Multiple classifiers" is where several classifiers are combined to identify attacks. "DTC4.5" functions as a signature identification module and "SVM" functions as an "anomaly detection module". The benefit of multiple classifiers-based approach is its lower false alarm rate and high detection rate.

Further investigation into standard feature selection has been carried for unknown attacks. The primary objective of this research paper is to complete a detailed evaluation and critical analysis of utilizing machine learning-based IDS/IPS in a specific wired network. Furthermore, concept inspection of numerous machines learning techniques have been completed and observations have been given in-relation each of its categories.

The chapter has been divided into four aspects:

- The categories of attacks founded upon their characteristics in relations to the low-frequency detection rate.
- The discussion of numerous pre-existing literatures for "intrusion detection" has been given.
- Critical analysis of different "intrusion detection" techniques is implemented in-relation to their attack types.
- In Conclusion future improvement for machine learning-based IDS/IPS to administrate.

2   Literature Review

There are many kinds of literature available for application for machine learning to IDS/IPS [6] have displayed the use of different data mining methodologies for anomaly intrusion detection". The author has classified the "anomaly detection" pillars "clustering-based approach, classification based and hybrid approaches. For example, K-means, K-Meoids, EM clustering and Outliers detection algorithms have been selected under cluster-based approaches. Naive Bayes, Genetic algorithm, Neural networks, support vector machines categorized under classification machine learning approaches. The author has given a contrast of papers applying the overarching evaluation method".

Ogino [7] exhibits that they are three different categories "supervised, unsupervised and reinforcement learning algorithms". In "supervised learning a classifier is educated on the labelled data set". "Unsupervised learning" is employed because there are no "labelled datasets". "Reinforced learning" when the IT network managers can label the "unlabeled instances".

Ahmed et al. [8] illustrates that on "network anomaly detection approaches. Intrusion is categorized into four classifications. "DoS, Probe, U2R, R2L based on KDD99 dataset". Each classification points to certain types of machine learning approaches. For example, categorization, clustering, statistical and information theory-based approaches".

Meryem and Ouahidi [9] entails that there are many similarities between "ML machine learning" and "DM data mining" as it practices the same methodologies for identifying intrusions. The author further emphasizes that KDD99 and DARPA data sets were used to train the algorithm. The author also explains that NetFlow and tcpdump dataset can be implemented for training as well.

In this chapter, a detailed investigation and evaluation of different machine learning techniques have been performed to determine the limitation related to identifying intrusions within networks. The use of different intrusion detection methods for different attack sets has been used. Emphasis on different factors while algorithm selection has been made. Attack classification has been given with attack examples about its attack features. Furthermore, low frequency and detection methodologies have been implemented.

The outline of distinct intrusion detection/prevention proposition is examined including the literature on distinct "datasets". I have found that traditionally researchers employ "KDD99 and DARPA datasets". Pre-existing intrusion detection practices based on machine learning capabilities have been carefully studied about its independent attack divisions with its constraints and its possible solutions. Further directions in the operation of deep learning and reinforcement learning for intrusion detection and prevention has been reviewed.

2.1 Classification of Attacks with Relevant Attack Preventions

Intruders exploit vulnerabilities available in the network/device by manipulating certain tools for example "Nmap, scapy, Metasploit, Armitage, Dnsiff, Tcpdump, Net2pcap, Snoop, Ettercap, Nstreams, Argus, Karpski, Ethereal, Amap, Vmap, TTLscan and Paketto" etc. For a secure network, both network and host-based protection are essential. (Host-based attack is when attacked attempts to bypass the security protocol in place and attacks the host machine directly).

In this section, we have entailed assaults which are categorized into three classifications. In each classification, we have also defined the important attack features for each "attack based on KDD99 dataset and UNSW-NB data set".

- DOS (Denial of service attacks):

  The DOS attack is "unavailability of service to the users" thus the name "DoS (Denial of service attack)". For example, if an attacker were to transmit numerous data packets/requests to a web server. Then the web server will be flooded with requests and will be able to reply to any other service request.

  About DoS, there is a DDoS attack in which multiple machines or "zombie" machines are used to transmit multiple requests to one specific server.

  DoS attack is categorized in the types of Bandwidth depletion and Resource depletion attacks. In a bandwidth depletion attack, the intruder attempts to overwhelm the network by data packets. Bandwidth depletion attack is further divided in the two distributions "Flooding and amplification attacks".

  Flooding attacks: the infiltrator will attempt to flood the network with any number of packets. "Amplification attack" where an "IP address" is smurfed from a victim machine to launch a "DDoS" attack to the network. Resource Depletion The intruder would tie up the resources of the network by exploiting the "network protocol such as Neptune and mail bomb". Some attacks have been explained below briefly. Land: This attack, an intruder transmits "spoofed" "SYN packet" where the "source address is the same as "destination address".

  Attack mitigations: These intrusions can be identified by utilizing the feature "Land". If the value of the feature is 1 then source and destination address are equivalent.

Mailbomb: An intruder would transmit numerous emails with attachments to a distinct "mail server" rustling the server to go offline.

Attack mitigations: This attack can be identified by reviewing thousands of emails being transmitted from one source address within a short amount of time. Points such as "Destination IP, total bytes, protocol and destination host address are used to detect this category of an attack.

- Scanning attacks

The iúntruder can scan devices in the network to gather information on the node info such as ("IP address, port and version scanning") in the network before executing a dedicated attack on vulnerable devices. An intruder transmits a large number of data packets to retrieve detailed information about the network/machine practicing tools such as "Nmap, satan, saint, msscan etc".

Reddy [10] exemplifies an informed analysis on scanning methodologies. In three parts Nature, Strategy and approach. The behaviors of the scanning attack can be passive and active.

An example of an attack and prevention is outlined below.

"Ipsweep": This attack is used to establish which devices are listening on the network by circulating many ping packets. If a target device replies it would define that the target device is alive and can be attacked.

Attack prevention: A NISDS or "Network Intrusion Detection System" inspect the total number of ICMP/ping packets within a certain time frame. For example, "duration", "protocols ("ICMP"), Dst host SRV rate (used to find the connection to the same service) and flags". Can be used for "ipsweep" detection.

- User to Root Attacks:

This attack is a combination of exploits which are operated to gain administrative access to a server via the standard user. For example, in "buffer overflow," the intruder manipulates the vulnerability of an application to transfer a high amount of data to a static buffer without confirming the data capacity. An intruder would employ the overflowing data to produce native commands completed by the OS. In "Ffbconfig" the intruder exploits the "ffbconfig" program allocated in some OSs. Infiltrator overwrites the "internal stack" of the program which does have a capacity checker.

Attack prevention: KDD99 dataset can be used to observe the behaviors of these types of attacks. For example, it has behavioral patterns such as "Num failed login", Su attempted" and "Is hot login" can be used to distinguish between normal and abnormal behaviors.

"Fuzzers": In this, the intruder transmits randomly generated large input patterns from the "command line" or in a "protocol packet form". Intruder attempts to identify security vulnerabilities in the "OS", "application" or "network". This is to make a resource unlivable for a certain time frame or crash it.

Attack Prevention: If a device is transmitting data packets in the large quantity continually through the same "service protocol or the port" during a certain period. This could be a "fuzzer" attack The prevention for this attack could be "sybites or spkts".

Several attacks have been illustrated in the section. Each attack is executed in a method and conveys its personality which has been given. The network prevention methods are a requirement for the detection of the attack categories discussed above. KD99 are traditional datasets which are used most commonly which have been considered for this literature review.

2.2   Machine Learnings Algorithms and Feature Selection

In this section, numerous and mostly used machine learning algorithms utilized for intrusion detection and prevention. This algorithm entails various behaviors and generates different results in intrusion detection. In this, we have discussed the functionality of different machine learning algorithms and their behaviors patterns including their pros and cons for the correct machine learning algorithm selection.

A.   Algorithms employed in Machine Learning

Machine learning algorithm practiced in two stages "training and testing". In the training stage, mathematical calculations are completed on training datasets and "learn the behaviors traffic over a period". In the testing phase, "test instance" is categorized as "normal" or "intrusion" behaviors. Numerous machines learning algorithms are given below:

(1)   "Decision Tree":

"Decision tree," an intrusion method, is practiced to display every possible decision/outcome. The learned patterns/trees are then transitioned into if-else rules. "Decision node" where each branch illustrates one of the possibilities of these tests. "Leaf node" depicts "the class to which the object belongs to". Several algorithms come in the "Decision tree" umbrella.

Three most popular algorithms are "ID3, C4.5, Cart and LMT tree". "ID3" illustrated algorithm employs "information gain criteria" Furthermore "ID3" is not able to manage "missing values and numeric" where "C4.5" algorithm is an improved version of it. "Logistic model tree" (LMT) operates a "decision" tree "linear regression model".

Decision tree operates on two methodologies to select best attributes "Entropy and Information gain". "Entropy characterizes the impurity of an arbitrary collection of examples whereas Information gain measures how well a given attribute separates the training examples according to their target classification".

Decision trees are applicable in situations such as "instances" can be identified by "attribute-value pairs" and every value can be separated into possible values, the out must be either yes or no, "data set" can have errors and missing values as the algorithm is robust to it.

The performance analysis of the decision tree has been carried out in the next section. "Decision tree" is more effective than other algorithms in "single classifiers".

The information gain function effective in the selection of the right values depends on how much information it was given. The top/points from which the tree splits are the critical aspect of a dataset.

They cannot identify any expectations. Furthermore, the maintenance cost for computing all the possible outcomes/possibilities is very high. The algorithm is not able to withstand data set modifications and it would deliver errors.

(2)   "Artificial Neural Network":

"Human brain" deciphers real-world situations and its context in a method where computers are not able to. "Neural network" was first introduced in the 1950s as a resolution to this problem. "Artificial neural network" is an attempted replication of the "network of neurons" that constructs the human brain so that the computer can "learn and make decisions" as a human would. "ANN" is generated by "programming machines" inmates and connect as brain cells would [5].

There are few examples of ANN Multilayer Perceptrons Back Propagation Algorithm, Adaptive Resonance Theory-based, Radial Basis Functions based, Hopfields Networks and Neural Tree. "ANN" contains three main aspects: input node, hidden nodes and output node.

"ANN" is a "nonlinear model" which can easily be employed. The detection of ANN for "low frequency" attacks. The detection is low. The training time frame is longer due to the "non-linear mapping of global approximation". There is a difficulty in identifying the number of neurons/devices and the hidden layers". Furthermore, required training dataset size could be large as well in relevance to the training.

(3)   "Support Vector Machine":

"SVM" is one highly utilized machine learning methodology for IDS/IPS. "SVM" is based on a separation of "two data classes". The error margin can decrease and increase the distance between "hyperplane and instances on either side of it". "Data points" that are found on the margin of "optimum separating hyperplane" have the name of support vector points and the result is given as a "linear combination of these points". SVM is not able to distinguish the "hyperplanes" separation and the resolution to this is "mapping the data into hyperplane space" known as "feature space". Illustrate the partition there. The Kernel function is utilized in data mapping for feature classification [5].

The "SVM" algorithm" kernel functions" has two subcategories "Linear and Nonlinear". "Linear SVM", the dataset ("training data") the data is linearly and if it is not then "non-linear SVM classifiers" produces poor results. Thus "nonlinear kernel" matches the data to a "higher dimensional feature space" to discover the "linear plane".

SVM can further be classified into type "Multi-class SVM" and "one-class SVM". "Multi-class SVM is employed for "supervised machine learning" In which it would employ a "set of binary SVM classifiers".

"One class is used for unsupervised machine learning algorithms" [5].

There are many disadvantages to SVM, the "high memory require-ment and algorithmic complexity". The performance is dependent on the "kernel function and its parameters" selection. "Linear SVM" results are inaccurate and overfitting. Training time longer than other algorithms which are not adequate as the user behaviors keep on changing [5].

(4)   "K-Means Clustering":

"K-means" algorithm is a "clustering-based anomaly detection algo-rithm". The foundation of this algorithm is "assumptions that normal data instances lie to their cluster centroid on the other hand anomalies are placed far away from it it's cluster". The other "data points" are designated to the "cluster" based on the distance to its "centroids". The next stage is "centroid" is calculated as average data points for each cluster. This is a continual process until "there is no change in centroid" as given in appendix 11. "K means clustering" has been widely adopted throughout the industry in IDS/IPS to improve its detection capabilities [5].

This methodology does not function if the data anomalies form a cluster by themselves. In this situation, this algorithm would not be able to identify instructions.

In this subsection, several machine learning algorithms were discussed such as supervised (SVM) and unsupervised (Kmeans clustering). The supervised machine learning algorithms are highly proficient in identi-fying known intrusions/attack behaviors and its labelled attack dataset. On the other hand, one-class SVM and K-Means clustering are unsupervised machine learning algorithms. Unsupervised machine learning algorithms are beneficial for identifying unknown/ zero attacks and discovering "out-liers". The data from the outliers can be extracted and used to find "distinct characteristics from data".

B.   "Feature Selection in Machine Learning"

The performance of the classifiers is founded upon two main aspects: "Classi-fier's technique and select feature subset". The main goal of "feature selection" is to confirm an optimal "subset of features". This action enhances perfor-mance, decreases resource/computing cost of the "classifier" and increases the efficiency of intrusion detection [5].

On the other hand, there are several drawbacks in considering all features in detection techniques such as, increase in computing/resource overhead and increase the training and testing time frame. The storage required is also higher as the number features in the database grows the more space is needed to deposit. It hinders the "generalization capability of a classifier" which employs "data mining techniques" for intrusion detection and prevention. It inflates the error rate of the "classifier" as irrelevant features disrupt the identification ability of relevant features [5].

The feature selection methodologies are divided into three aspects "filter methods, wrapper method and embedded methods. The definition of these has been given in appendix 12. Filter technique does not include classifiers. The data computing and processing is faster than other methods. The disadvantage of this method is that performance of other classifiers is not considered, thus failing "to give the best feature subset for classification." the disadvantage of "wrapper based" continuous learning could result in overfitting the problem. It is very resource/computation intensive as it would require continuous trails to discover the successful algorithm and classifier sub-set iteration. "Embedded is combined with the classifier's design in the training phase. Data exploitation is optimized which will reduce the number of retraining of the classifier for each new subset. Hybrid methods have higher computational cost than filter-based methods." [5].

Filter methods	Wrapper methods	Embedded methods
Generic set of methods which do not incorporate **a specific machine learning algorithm**	Evaluate on a **specific machine learning algorithm** to find optimal features	Embeds (fix) features during **model building process.** Feature selection is done by observing each iteration of model training phase
Much **faster** compared to wrapper methods in terms of time complexity	**High computation time** for a dataset with many features	Sits **between filter methods and wrapper methods** in terms of time complexity
Less prone to **over-fitting**	High chances of **over fitting** because it involves training of machine learning models with different combination of features	Generally used to reduce **over-fitting** by **penalizing** the coefficient of a model being too large
Examples—**Correlation, Chi-square test, ANOVA, information gain** etc.	Examples—**Forward selection, backward elimination, stepwise selection** etc.	Examples—**LASSO, Elastic net, Ridge regression** etc.

In this section, the number of machines learning algorithms such as "decision tree, artificial neural network, SVM and K-means clustering overview" have expanded upon. The characteristics, advantages and disadvantages have been provided. The importance of feature selection and various feature selection has been demonstrated these can be employed with the machine learning algorithm, Feature selection allowed me to discover "important, non-redundant and relevant attributes that aide in the accuracy of the "predictive model" Furthermore it enables me to distinguish between relevant and irrelevant features to improve the performance of the classifiers. Furthermore, by reviewing different machine learning algorithms the best selection would be SVM as it is best optimized for detecting zero-day attacks [5].

2.3 Data Mining Tools for Machine Learning

There many tools that pre-exist within the market that incorporate numerous machines learning algorithms for which a number of them are listed below.

A. "Weka"

"Waikato Environment of knowledge analysis (Weka) is a machine learning tool produced by University of Waikato, New Zealand in 1993, This is an open-source tool written in Java language. This tool has different functions such as "data processing, feature selection, classification, clustering, regression and visualization" (Weka 3—Data Mining with Open-Source Machine Learning Software in Java 2020).

B. "Scikit-Learn"

"Scikit-learn" is an "open-source machine learning library" produced by Google". This is written in python and can provide several ml tools such as classification, clustering, regression algorithms"

C. "TensorFlow":

"Tensor Flow" is open-source software for ML produced by "Google brain team". It was covered by apache 2.0 open-source license in Nov 2015. "TensorFlow" is a productive tool adapted for deep learning as it is used for "building and training neural networks". "Data flow" graphs are operated to construct ML models and action computations. "Data arrays are edges between nodes of graphs" are known as "tensors". "TensorFlow" functional multiple APIs, for example, C++, Go, Java, Haskell and Rust APIs and it supports multi-CPU/GPUs.

In the section, some of the data mining tools were discussed and some of these tools can also support deep learning algorithms. Many of these tools have GUI's interfaces for high accessibility.

3 Methodology and Data Analysis

3.1 Software Development Methodologies

The software development life cycle is a composite of foundational eight pillars that followed every software/product creation. Given below:

1. "Conception
2. Requirements gathering/exploration/modeling
3. Design
4. Coding and debugging

5. Testing
6. Release
7. Maintenance/software evolution.
8. Retirement" [11].

There are different software development life cycles/methodologies given below for software development and production.

3.1.1 "Waterfall"

The "waterfall" is a traditional and a plan-driven process model.

The "waterfall model" is a "linear sequential flow". The progression of software development moves in a downward slope (relation to a waterfall). It means the passage to the next phase in development procedure commences after completing the previous stage. This model does not delineate a process to step back to the last degree to administer changes in requirements [11].

There are many different advantages and disadvantages for the application of this model:

Advantages

• Accessibility and ease of usability
• Structured approach
• Stages and activities are defined beforehand.

Disadvantages

• Can changes the requirements or go back a step/ a phase
• Costly and time exhaustive.

3.1.2 "Agile Model."

Agile model is an evolved version of an "iterative model", where requirements and solutions matured through "cross-functional" teams.

There are many different advantages and disadvantages for the application of this model:

Advantages

• Adapt to system failures and requirement change
• Better client/stakeholder communication and continuous input from the client
• High-quality product releases with a given time frame.

Disadvantages

• Scalability
• Require special training and skills.

3.1.3 "Iterative Model"

The "iterative model" is an evolved version of the waterfall to reduce the risk. It proceeds with "initial planning" and completes with deployment in continuum interactions in the middle. Its foundation is to produce a system through reciprocated iterations and in a smaller portion at a time.

3.2 Software Development Methodology Selection and Algorithm Creation Guide

This section has been split into two sub-section software development methodology selections with its limitations and with its mitigations. Moreover, the Algorithm creation guide has been discussed within this section.

3.2.1 Software Development Methodology Selection

The author has developed ML practicing Agile software development because of its nature, consent change in requirements and robustness. The project was to produce an improved version of their current IDS/IPSs for the client "Pilot cybernetics" with ML (Machine learning).

In creation of the ML (Machine learning) algorithm, application of Agile methodology is imperative as the algorithm partitioned into sprints/phases, which required it's testing and documentation completion. The author has further illustrated in the "manual for algorithm creation section where the development of the algorithm has partitioned into sprints.

There are few limitations associated to the Agile software development methodologies associated to it, but their mitigations that applied it to overcome it as given below:

Limitations of Agile

There are five critical limitations to selected (agile methodology) which are given below:

(1) "Poor resource planning":

Very little resource planning is applied within the agile method as the team or organization does not know what the results will look like or have its idea. It is very challenging to predict cost, time, or resource required at the start of the project [12].

(2) "Limited documentation":

The documentation is created within agile development is very limited, as the documentation is generated throughout the project and not at the start. There is significantly less detailed and incomplete documentation produced [12].

(3) "Fragmented result":

The incremental delivery can produce a product quickly. Still, there is also a downside to it as different teams establish components in various iterations the complete product can arrive in fragments rather than the entire unit [12].

(4) "No finite end":

There is an end to the product as there is minimum planning in the earlier stages the project can get sidetracked in designing the other functionalities of the output [12].

(5) "Difficult Measurement":

The project performance measurement is difficult as it would require KPI that needs to be placed across the iterations, but the nature of agile is delivery increments. It causes challenges in measuring the performance of the entire project [12].

Mitigations

There are few limitations to this methodology's application, implementing the below resolution the Agile methods can be utilized for this project:

- We are delivering a quality end product instead of providing a functioning product to the client [12]. The above point can be administered to this ML algorithm project development by Harding and pre-defining the project's requirements first and setting up train data for the ML to learn from so it would be able to detect unknown attacks more effectively [12].
- The team should have a predefined and transparent process in delivery for the product [12]. This can be executed for the current project by setting a strict ML development process such as developing the training data and then the ML module to follow [12].

The second subsection contains a full instructional manual given the algorithm's creation and the application of agile methodology with the algorithm's development.

3.2.2 Step by Step Guide to Create the Machine Learning Algorithm

This manual defines the creation and testing of the machine learning algorithm. Author has trained the model on different machine learning algorithms, compared the accuracy for all of them and selected which algorithm would work the best.

Below are the variables/tools used for algorithm creation.

- Python (Jupiter) to create the algorithm in

- Test data and learning data from the client current network info, survey monkey gather data.

Product backlog

A product backlog created consisting of all the significant tasks model creation, learning data creation, test data creation, model creation and evaluation.

Author has developed the algorithm in 6 sprints and it was produced from the product backlog tasks with its own sprint logs. As shown below.

First Sprint

The first sprint's backlog incorporates the learning and test data from the clients' current network.

The author also created a questionnaire and examined the client's network transmission to understand the connections and features running over the network.

Second Sprint

In this sprint, pie-chart and bar charts were created for visual aid purposes to determine whether a network was affected by a yes or no cyberattack.

The author has counted how many rows are YES & how many of them are NO.

The code below can generate a bar chart from the data in the training file.

Distribution Cyber Attack (YES/NO) Plots

```
count_Class=pd.value_counts(train["Cyber Attack"], sort= True)
count_Class.plot(kind= 'bar', color= ["blue", "orange"])
plt.title('Bar chart')
plt.show()
```

Once the second sprint had completed. The produced solution was presented to the client to gain the feedback and the client was content with the solution produced.

Third Sprint

Feature selection is a critical task in the world of Machine Learning as there are about **40+** columns in our dataset. To classify the essential elements from un-important to determine whether an attack has been conducted or not.

The author applied the code below essential features in the dataset.

```
9]:  from sklearn.ensemble import RandomForestClassifier
     rfc = RandomForestClassifier();

     # fit random forest classifier on the training set
     rfc.fit(train_x, train_y);
     # extract important features
     score = np.round(rfc.feature_importances_,3)
     importances = pd.DataFrame({'feature':train_x.columns,'importance':score})
     importances = importances.sort_values('importance',ascending=False).set_index(
     # plot importances
     plt.rcParams['figure.figsize'] = (11, 4)
     importances.plot.bar();
```

If all the features are utilized for the model, the ML algorithm will be resource and time exhaustive, so only the required elements must be adopted.

After applying some coding techniques on our dataset, the Author found the following essential features of our dataset.

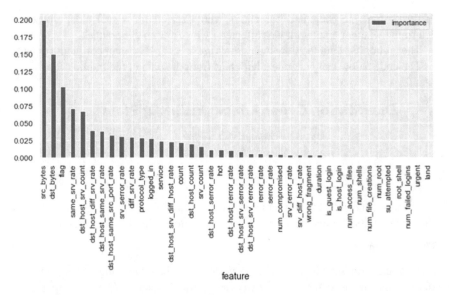

The image can distinguish the required and not-required features. The above plot shows the importance of every element.

Once the Third sprint had completed. The produced solution was presented to the client to gain the feedback from him and the client was content with the solution produced.

Forth Sprint

The purpose of this sprint was to find the top most needed features and neglect all others. After using the code given below, the Author can view essential elements to identify attacks.

These features will be fed into the prediction model to predict whether an attack has been conducted within our network, applying the test data.

```
: from sklearn.feature_selection import RFE
  import itertools
  rfc = RandomForestClassifier()

  # create the RFE model and select 10 attributes
  rfe = RFE(rfc, n_features_to_select=15)
  rfe = rfe.fit(train_x, train_y)

  # summarize the selection of the attributes
  feature_map = [(i, v) for i, v in itertools.zip_longest(rfe.get_support(), tra
  selected_features = [v for i, v in feature_map if i==True]

  selected_features
```

```
: ['src_bytes',
   'dst_bytes',
   'logged_in',
   'count',
   'srv_count',
   'same_srv_rate',
   'diff_srv_rate',
   'dst_host_srv_count',
   'dst_host_same_srv_rate',
   'dst_host_diff_srv_rate',
   'dst_host_same_src_port_rate',
   'dst_host_srv_diff_host_rate',
   'protocol_type',
   'service',
   'flag']
```

This code was developed with the importing "itertools" module to retrieve the most necessary features and store in the array "feature_map.

Fifth Sprint

Flowchart to display the process. Furthermore, it elaborates on how classifiers will look like and can be trained to predict results.

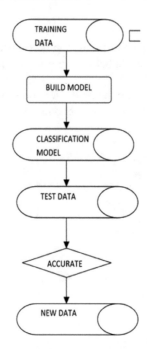

Before creating the model, data set was splintered into the train and test set, I have split the data in 30% testing set and 70% for the training set.

The task was completed using the code given below.

DATASET PARTITION

```
1]:  from sklearn.model_selection import train_test_split
     X_train,X_test,Y_train,Y_test = train_test_split(train_x,train_y,train_size=0.70, random_state=2)
```

Sixth Sprint: Training the Modules

Author generated a model on a different machine learning algorithm, and I have used a comparison point select an ML algorithm which provides the high accuracy in detecting attacks. Given below:

- K-nearest neighbors
- Logistic regression
- Gaussian Naïve Bayes
- Decision tree.

The author has used training data to determine the detection accuracy and train to the ML modules to identify unknown network intrusions that could occur within the client's network and have discovered that "decision tree" can produce the highest detection as displayed in Table 1.

Table 1 .

	Training accuracy	Confusion matrix
Naïve Bayes	0.907	[[8997 392] [1245 7000]]
Decision tree	1.0	[[8997 392] [1245 7000]]
K-nearest neighbors	0.993	[[9356 33] [77 8168]]
Logistic regression	0.954	[[9076 313] [482 7763]]

"Decision tree" algorithm is part of the supervised learning algorithm umbrella, but the "decision tree" can be applied to resolving "regression and classification problems" [13].

To gain the above results, the author had to develop the code given below.

EVALUATE MODELS

```
In [71]: from sklearn import metrics

models = []
models.append(('Naive Baye Classifier', BNB_Classifier))
models.append(('Decision Tree Classifier', DTC_Classifier))
models.append(('KNeighborsClassifier', KNN_Classifier))
models.append(('LogisticRegression', LGR_Classifier))

for i, v in models:
    scores = cross_val_score(v, X_train, Y_train, cv=10)
    accuracy = metrics.accuracy_score(Y_train, v.predict(X_train))
    confusion_matrix = metrics.confusion_matrix(Y_train, v.predict(X_train))
    classification = metrics.classification_report(Y_train, v.predict(X_train))
    print()
    print('============================== {} Model Evaluation =============================='.format(i))
    print()
    print ("Cross Validation Mean Score:" "\n", scores.mean())
    print()
    print ("Model Accuracy:" "\n", accuracy)
    print()
    print("Confusion matrix:" "\n", confusion_matrix)
    print()
    print("Classification report:" "\n", classification)
    print()
```

The testing results and accuracy result delivered to the client once the training module sprint was completed to retrieve feedback from the client and in the feedback the client has stated due to the budget restraints and time restraints. The client requires the algorithm to detect the unknown and zero-day attacks and not prevent them and would consider the current solution as complete. This has been further illustrated in Appendix 7.

Testing in the client's networks

Once the decision was made that the "decision tree" has the highest detection rate, this algorithm was then implemented within the clients' network, and it was tested (live environment). This was done to determine that the ML algorithm can detect zero-day attacks if they exist within the client's network. Thus, given below.

The table demonstrates the accuracy matrix of the different ML modules within the client's network. It clearly shows that the "decision tree" has the highest accuracy detection rate within the anonymous data and instructions.

Table 2 .

	Testing accuracy	Confusion matrix
Naïve Bayes	0.906	[[3872 188] [517 2981]]
Decision tree	0.994	[[4035 25] [15 3483]]
K-nearest neighbors	0.991	[[4037 23] [40 3458]]
Logistic regression	0.955	[[3922 138] [201 3297]]

Evidence

Furthermore, "decision tree." has detected more network intrusions than other ML modules within the live environment, as it has detected 11,731 instructions in relation to other ML algorithm modules where K-Nearest Neighbors had 8930, Naive Baye had 8284 and logistic regression module had 9170.

```
In [34]:  # PREDICTING FOR TEST DATA using KNN
          pred_knn = KNN_Classifier.predict(test_df)
          pred_NB = BNB_Classifier.predict(test_df)
          pred_log = LGR_Classifier.predict(test_df)
          pred_dt = DTC_Classifier.predict(test_df)

          #counting how are their yes and how are their no's
          # KNeighborsClassifier Model Test Results
          algorithms = ['KNeighborsClassifier','Naive Baye Classifier','LogisticRegression','Decision Tree Classifier']

          countY = 0
          countN = 0

          for i in range(len(pred_knn)):
              temp = pred_knn[i]
              if temp == "YES":
                  countY += 1
              else:
                  countN += 1
          print()
          print('============================== '+algorithms[0] +' Model Test Results =============================='.format(i))
          print()
          print ("Number of intrutions detected", countY)
          print ("Number of safe connection", countN)
          print ("The total number of connections" , countY+countN)
          countY = 0
          countN = 0
          i=0
```

The above image exhibits the code which was used to retrieve the matrix for network intrusion detection.

Hence, the author has successfully utilized the machine learning against the cyber-attacks and uncovered that the "decision tree" algorithm produces the highest result from training data/testing data and once established in the client's network.

4   Critical Discussions

This section contains the evaluation of the ML algorithm that has been produced for the client and concentrate upon the objective of the ML algorithm and its application. It is then evaluating the currently produced results.

The research study's primary purpose is to produce an ML algorithm applied to IDS/IPS to identify and prevent unknown/zero-day attacks. Thus, improving traditional signature-based IDS/IPS' effectiveness and it's efficiency. This project was commissioned by the client "Pilot cybernetics ltd" to enhance their current solution of IDS/IPS.

The key objectives and goals/scope for the product is listed below.

Goals

- Research will concentrate on small but high-security threats to networks which are zero-day attacks.
- The project will attempt to upgrade the traditional IPS/IDS of "Pilot Cybernetics" from "Signature-based" to "Anomaly-based."
- The IDS/IPS will operate unsupervised machine learning algorithms such as K-Means clustering.

Objectives

- Deliver, implement and test upgraded versions of IPS/IDS with machine learning algorithms to fit the client's requirements.
- Investigate and research current machine learning algorithms such as K-Means clustering and IPS intrusion systems such as SNORT to be applied to the upgraded IPS/IDS.
- Develop a new algorithm or fusion of current machine learning algorithms applied to the upgraded IPS/IDS. To resolve, detect and prevent a 0-day attack for the client.
- Full testing will be completed on the newly upgraded IPS/IDS to test whether they can detect zero-day attacks or not.
- Testing should be completed both in test and live environment.

One of the key objectives was to produce an ML algorithm for their current IDS/IPS, which would detect and prevent zero-day and unknown attacks for the client.

The objective was met to a certain extent where the ML algorithm was created for IDS, where it would be able to detect any unknown and zero-day attack that would occur in the network. However, it would not be able to prevent those attacks. The author could not meet the objective due to the change in requirement from the client and time constraints.

The client has requested for the requirements for the project to be amended as they only require the algorithm to detect unknown/zero-day attacks and not prevent them. This was due to time constraints and budget constraints.

As to create an ML algorithm which would detect and prevent unknown attacks is costly, to which client opted for a more economical solution. Furthermore, only the project objective was not met due to lack of time and cost restriction. However, if there was more time and if there was no budget constraint then a better solution could have been presented.

The current solution has a detection module which is able to detect any unknown and zero-day attacks, to which an accuracy matrix has been created to view how many detects a specific module can detect and what accuracy rate. For results that were gained was that the decision tree has the highest detection rate and was able to detect the highest number of unknown attacks made within the network.

The solution is able to meet the other objectives of research such as delivery, testing and implementation of IDS/IPS with an application of ML algorithms. An ML-based IDS/IPS system was produced with correct testing and implementation in the client's network.

This was further discussed in above sections, but in essence, the first test data was gathered from the client network to test the ML module in a controlled test environment. Furthermore, the same data was also used to train the module to detect unknown attacks. Once the ML module was tested and trained with test data. It was then implemented in the client's network and tested again.

The ML algorithm was created using the Agile software development methodology as it is best suited for developing the current solution. For example, the current solution has an aspect where the requirements of the solution can vary from the start of the project, and the client would like to have their input while the development of the ML algorithm.

There are few limitations in selected methodology; such as lack of documentation, but this can be mitigated as if the process and procedure plan is created for the development, then a meaningful documentation could be generated once the product is produced.

The author has measured the end product in two separate entities which are qualitative and quantitative. Qualitative measurement (This is when the product or results are measured by quality such as description and feelings) and quantitative measurement/analysis (This is when the product is analyzed using numbers and data) [14].

In Quantitative analysis user stories and surveys were used to analyses the completed solution. On the other hand, in qualitative graphs and detection matrices were utilized.

The qualitative measurement of the product made is from the user stories. The current solution produced meets the first 2 user stories criteria. ML algorithm can provide both detection accuracy and the number of intrusions made against the network.

Furthermore, concerning user stories 3 and 4, the algorithm can generate a bar chart and pie chart for the client to visualize the intrusion detected in a visual format. The client completed more over the final survey/questionnaire to gain the last feedback about the finished product and product implementation.

These can be operated to analyses and measure the product results in quantitative analysis as the current solution suffices both the user stories requirements and feedback received from the client.

The solution produced was also measured/analyzed in quantitative analysis. To which the detection matrix (ML detection accuracy in detecting unknown attacks) is investigated. There was a clear distinction that "decision tree" was able to outperform

Table 3 .

	Training accuracy	Confusion matrix
Naïve Bayes	0.907	[[8997 392] [1245 7000]]
Decision tree	1.0	[[8997 392] [1245 7000]]
K-nearest neighbors	0.993	[[9356 33] [[77 8168]]
Logistic regression	0.954	[[9076 313] [[482 7763]]

all the other ML algorithm modules as it had the highest detection accuracy results and was able to detect the most increased number intrusion made against the network.

The analysis of the results created by the ML algorithm/solution demonstrates that the "decision tree "algorithm was able to outperform all the ML modules used for detection. The table below displays the accuracy detection when the algorithm was being trained with the test data set (Table 3).

The above table explicatively demonstrates that the "decision tree" has the highest detection accuracy than any other ML module. The "decision tree" accuracy rates 1.0, on the other hand ML modules such as "K-Nearest Neighbors" has the detection rate of 0.993.

The same can implied for the results in the live environment where the "decision tree" algorithm is able detect highest number of intrusions with in the network and has highest detection accuracy as depicted in the Table 4.

The above table explicatively demonstrates that the "decision tree" has the highest detection accuracy then any other ML module. The "decision tree" accuracy rates 0.994, on the other hand ML modules such as "K-Nearest Neighbors" has the detection rate of 0.991.

This algorithm can be further improved with larger data scope as the current data scope is very limited due size of the corporation. The number of nodes and transmission completed with its network. However, if the training data and implementation was performed in a large organization then data scope would be larger thus the accuracy would have further increased.

Table 4 .

	Testing accuracy	Confusion matrix
Naïve Bayes	0.906	[[3872 188] [[517 2981]]
Decision tree	0.994	[[4035 25] [[15 3483]]
K-nearest neighbors	0.991	[[4037 23] [[40 3458]]
Logistic regression	0.955	[[3922 138] [[201 3297]]

The algorithm that has been produced can be a very beneficial academic body as it is able to provide a comparison between a number of ML algorithms such as "Naïve Bayes", "Decision Tree", "K-Nearest Neighbors" and "Logistic Regression". For this a table was created which showed the detection accuracy for both aspects one for training the algorithm and testing the algorithm. The table has displayed Tables 1 and 2. The Table 1 demonstrates that the "decision tree" has the highest detection accuracy with the training data set, consisting of transmission data gathered from the client. Also, in Table 2 the same algorithm "decision tree" had a higher detection rate compared to other ML modules when it was implemented within the client's network and there were unknown instructions.

On the other hand [6] have displayed the use of different data mining methodologies for anomaly intrusion detection". The author has classified the "anomaly detection" pillars "clustering-based approach, classification based and hybrid approached K-means, K-Meoids, EM clustering and Outliers detection.

The Algorithm that has been constructed for the duration of the project varies from the ML algorithm suggested by the author [6] as the author has suggested clustering based solution with K-means clustering. On the other hand, results reveal that K-means clustering delivers lower detection accuracy results in comparison to the "decision tree" ML algorithm.

The literature review was very useful and highly optimized throughout the research study. Furthermore, it was highly influential in the creation of the ML algorithm for the client.

The literature review has given the author an insight into what is related and pre-existing work that has been completed by other authors. For example, [6] has stated that different "data mining methodologies" can be applied to "anomaly-based" IDS/IPS.

The author has further noted that "anomaly-based" is founded upon "clustering-based", "classification based and hybrid approaches. For example, "K-means, K-Meoids, EM clustering and Outliers detection algorithm" are categorized as "cluster-based approaches". Moreover, there was an evaluation of these approaches in the author's papers.

This research is aimed at to understand the principles that "anomaly-based" IDS/IPS are based upon with its corresponding ML algorithms. Furthermore, these were later utilized in creating the detection module, which determines which ML algorithm module provides the highest detection accuracy.

The attack classifications and its prevention methods from the literature review were highly beneficial in the creation of the ML algorithms test and learning data. In the sub-section of the literature review, multiple attacks and intrusion were discussed with its features. For example, "user to root" attack would use specific features, for instance, the port 80 and number of bytes transmitted it could use would be high. For which all this information was considered when creating training data for the algorithms to practice and for testing it in the test environment.

The analysis of different machine learning algorithms available in the modern world was highly beneficial in developing/testing the algorithm and evaluating the completed solution.

In developing the algorithm, the information from the literature review was operated to understand what ML algorithms should be performed to create a detection module and which would have the best results.

Naïve Bayes, Decision Tree, K-Nearest Neighbors and Logistic Regression selected their high detection capabilities. Furthermore, "feature selection" methodologies such as filter method, wrapper method and the embedded method were advantageous in selecting the most essential features in the test dataset.

Many different techniques could have been applied for this study, but due to the time and resource constraints, they were ignored in the creation and implementation stage.

The first technique that was overlooked was the waterfall software development module when creating the ML algorithm. The author believed that the waterfall would not be a good fit for creating an algorithm, as in the waterfall model, there is no prospect of change in requirements or moving back to a previous phase. On the other hand, if the waterfall model were to be selected then the client end solution would have not been completed to its highest standards as the requirements of the project had changed while the solution was being developed.

The development of the ML algorithm required a software development methodology that is able to take in account the change in requirements. Consequently, the agile software development methodology was selected for that capacity. The second technique that was overlooked was selection of a variety of the machine learning algorithm when creating the detection module in the algorithm.

The "artificial neural network" algorithm is discarded in consideration for the detection module in ML solution. This was due its nature and algorithm functionality operation, as the algorithm functions with multiple applications/devices.

In contrast to the map of a human brain such as neurons firing and transmitting information. These devices function in cogent with each "learning and making decisions" from each other [5]. Furthermore, the algorithm was also neglected because of lack of resources and time as the author only had access to one PC for algorithm compute, thus this methodology was excluded in the development of the algorithm.

There were a number of other hypotheses that could have developed and tested in contrast to the current project aim. These have been given below:

Approach of applying ML in Firewalls.
Approach of applying block chain to cybersecurity environment.

- Approach of applying ML in Firewalls:
  Cooney [15] states that the "security attacks" are evolving in constant pace and signature-security components such as firewalls are not able to keep up the pace with it. The ML based firewall would be able to collect "telemetry information forms the network and compose it with pre-existing data sets, the firewall can "learn behaviors, recognize trends and recommend appropriate security policies". ML based firewalls would include stateful packet inspection with added ability to make decisions "based on application, user and content."

In this alternative hypothesis the author would compare the traditional and ML applied firewalls. Moreover, the author would have conducted a research study developing ML modules which detect, learn and suspend any malicious connections made towards the network.

- Approach of applying block chain to cybersecurity environment:
In this hypothesis the author would have discussed application of block chain networks in many different aspects of cybersecurity such as IoT devices [16]. Blockchain enables IoT devices to enable them to make decisions "without relying on the central authority". For example, "device can form a group consensus regarding the regular occurrences, within a given network" and close any "nodes" that function doubtfully.

There are few different types of evaluation criteria that could have been considered as follows.

Evaluation Guidelines [17] illustrates. "Connectedness": In these the evaluation criteria for this research study task completed to resolve issues for the shot turn could have been further discussed with the evaluation and how they set up there is permanent resolution rather than a quick work around.

"Coherence": In this evaluation of the solution the project could have taken in consideration for evaluating the solution in a more "humanitarian" light and took in employees of the client in the consideration as well.

The results from the testing of the solution and the results from the live solution were very reliable and complete as the algorithm was performed in both environments in a controlled test environment and client's live environment. As in the test environment the transmissions of the network can be simulated correctly to understand the capacity of the solution. Consequently, the algorithm should perform and provide accurate results, which it did.

The conclusions of each of the chapters are very reliable as they provide the brief summary of what has been completed with each of the chapter and how it all correlates to the continual study. For example, the conclusion in the literature review chapter demonstrates how the hypothesis/research question correlates and contrasts with the related works completed by the past authors.

5   Conclusions and Future Work

There are different recommendations that can be made for better performing ML application security components such as a different outlook could be taken on this project such as deep learning could be employed instead of a "decision tree", as it would be able to provide even more efficient result in intrusion detection and autonomous prevention. Furthermore, in future, an application of a GUI interface is highly required for better accessibility for the client.

In future work, this research is not sufficient as there is more improvement in intrusion detection and prevention is required. The traditional IDS/IPS must evolve to counter the threat of ever-evolving network attacks and application of deep learning

ML with security components such as firewall, IDS/IPS and sandboxes can be deployed to counter growing threats.

A "Deep learning" is an evolved version of neural networks. Deep learning employs multiple hierarchical information-processing layers for clarification and feature representation. The IDS/IPS efficiency and effectiveness can be further increased with the implementation of deep learning [18] illustrated that deep learning is split into three sub-categories "generative" ("unsupervised"), "discriminative (supervised)" and "hybrid". "Unsupervised"/"generative" deep learning operates following methods "Autoencoder (AE) and Boltzmann machine (BM). Comparable to "ANN", "AE" practices hidden layers, however, it operates three layers. The nodes in input layer and output layer equivalent. The nodes' operation purpose is to find the "feature set" aptitude. "BM" makes calculated decisions through "neuron's structure of binary units". Supervised learning recognizes some data parts and behavioral classifications for example Convolution Neural Network (CNN). Hybrid operates both models such as hybrid infrastructure of (DNN) "Deep neural network").

There are many challenges that have been encountered in the application of deep learning with intrusion detection and prevention within organization networks. The author [19] has attempted to administer deep learning with the IDS/IPS system as shown in appendix 13. This was done by the transmission of malware code into the image as input to learning attack behaviors. There are many advantages of utilizing deep learning contemplating the supervised and unsupervised deep learning approaches that advance the detection ability. Deep learning can adapt to data change as it executes a comprehensive data analysis. On the other hand, there are disadvantages in applying deep learning to security components such as it is very resource-intensive.

Deep learning is the future of ML as it can make the learning of huge collections of data more effective. It has been exercised for pattern classification and resource control over the past years. However, it can be applied to intrusion detection and prevention.

The threat of cyber-attacks such as zero-day attacks is on the rise. The network attack is an intrusion method where an attacker attempts to gain entry to the network in the intention of either stealing information or causing destruction within the network. For example, users to root attack. There traditional signature-based IDS/IPS as prevention/deterrent to these attacks, but these systems cannot identify zero-day or unknown attacks thus cannot stop.

This is because traditional IDS/IPS function on a signature-based principle to which it only detects the data packets or intrusions to the signature which is stored in the components database. The purpose research is to construct an anomaly-based IDS/IPS with application ML algorithm to which would be able to detect and prevent unknown and essentially zero-day attacks made against organization networks.

In this research evaluation of different network attacks and ML algorithms was completed to understand the core functionality of the algorithms. In listing the number of features a feature selection method is employed to determine important features, thus enabling the author to distinguish between yes and no. A comparison of the accuracy of various classifiers in detecting network intrusions attributes and it was selected the "decision tree" ML methodologies delivers the highest classification accuracy & lowest misclassification rate. Only 99% accuracy has been achieved at the current point, but it can increase if there is a large corpus of training data.

References

1. Fimia L (2020) Laughing all the way to the bank—DDoS bank cyber attacks on the rise. [Blog]. https://activereach.net/newsroom/blog/laughing-all-the-way-to-the-bank-ddos-att acks-on-the-rise/. Accessed 4 Oct 2020
2. Auld A (2020) Why has there been an increase in cyber security incidents during COVID-19?. [Blog]. PwC UK Cyber Threat Intelligence
3. Mishra P, Varadharajan V, Tupakula U, Pilli E (2019) A detailed investigation and analysis of using machine learning techniques for intrusion detection. IEEE Commun Surv Tutor 21(1):686–728
4. Software Reviews, Opinions, and Tips—DNSstuff (2020) What is an intrusion detection system? Definition, types, and tools—dnsstuff. https://www.dnsstuff.com/intrusion-detection-system. Accessed 7 Oct 2020
5. Razia S, Ramani Varanasi V (2019) Intrusion detection using machine learning and deep learning. Int J Recent Technol Eng 8(4):9704–9719
6. Yemunarane K, Hema A (2018) A survey on stress detection using data mining techniques. Int J Comput Sci Eng 06(08):27–29
7. Ogino T (2015) Evaluation of machine learning method for intrusion detection system on Jubatus. Int J Mach Learn Comput 5(2):137–141
8. Ahmed M, Naser Mahmood A, Hu J (2016) A survey of network anomaly detection techniques. J Netw Comput Appl 60:19–31
9. Meryem A, Ouahidi B (2020) Hybrid intrusion detection system using machine learning. Netw Secur 2020(5):8–19
10. PR, Reddy E (2018) A comprehensive survey on semantic based image retrieval systems for cyber forensics. Int J Comput Sci Eng 6(8):245–250
11. Dooley J (n.d.) Software development, design and coding
12. Lynn R (2021) Taming the Agile Chaos: who's who in your zoo?
13. Chauhan N (2020) Decision tree algorithm, explained—Kdnuggets. KDnuggets. https://www.kdnuggets.com/2020/01/decision-tree-algorithm-explained.html. Accessed 29 Dec 2020
14. SHIFT Communications—Integrated Communications + PR Agency—Boston | New York | San Francisco (2021) Understanding qualitative and quantitative analysis—SHIFT Communications—Integrated Communications + PR Agency—Boston | New York | San Francisco. https://www.shiftcomm.com/insights/understanding-qualitative-quantitative-analysis/#:~: text=Qualitative%20analysis%20fundamentally%20means%20to,its%20quality%20rather% 20than%20quantity.&text=Quantitative%20analysis%20is%20the%20opposite,%2C%20m easures%2C%20numbers%20and%20percentages. Accessed 5 Jan 2021
15. Cooney M (2020) Machine learning in Palo Alto firewalls adds new protection for IoT, containers [Blog]. https://www.networkworld.com/article/3562705/machine-learning-in-palo-alto-firewalls-adds-new-protection-for-iot-containers.html#:~:text=The%20machine% 20learning%20is%20built,%E2%80%93%20with%20behavior%2Dbased%20identification. Accessed 8 Jan 2021
16. Arnold A (2021) 4 promising use cases of blockchain in cybersecurity [Blog]. Forbes. https://www.forbes.com/sites/andrewarnold/2019/01/30/4-promising-use-cases-of-blockchain-in-cybersecurity/?sh=786acb493ac3. Accessed 9 Jan 2021
17. Netpublikationer.dk (2021) Evaluation guidelines. http://www.netpublikationer.dk/um/7571/html/chapter05.htm. Accessed 9 Jan 2021

18. Wani M, Khoshgoftaar T, Palade V (n.d.) Deep learning applications
19. Seok S, Kim H (2016) Visualized Malware classification based-on convolutional neural network. J Korea Inst Inf Secur Cryptol 26(1):197–208
20. Porter J (2020) Amazon says it mitigated the largest DDoS attack ever recorded. [Blog] An attack with a previously unseen volume of 2.3 Tbps. https://www.theverge.com/2020/6/18/21295337/amazon-aws-biggest-ddos-attack-ever-2-3-tbps-shield-github-netscout-arbor. Accessed 4 Oct 2020

Utilising Machine Learning Against Email Phishing to Detect Malicious Emails

Yogeshvar Singh Parmar and Hamid Jahankhani

Abstract Phishing is an identity theft evasion strategy used in which consumers accept bogus emails from fraudulent accounts that claim to belong to a legal and real company in the effort to steal sensitive information of the client. This act places many users' privacy at risk, and therefore researchers continue to work on identifying and improving current detection instruments. Classification is one of the machine learning methods that can be used to detect emails received. Different classification algorithms such as Naïve Bayes and Support Vector Machine (SVM) are discussed and compared in the course of this study. In an integration of the monitored and unregulated strategies, a new method has been developed to detect phishing emails. The research also contrasts the collection classes for manual and automatic emails. Series of terms are used to acquire words to differentiate between malicious and non-malicious communications in this research. In predicting the class attribute, the exactness of the different classifiers has been compared. SVM approach has the most reliable classification and misclassification rates of malicious emails than the Naïve Bayes method. To date, 98% precision was achieved, but if a researcher has a big corpus of training data, it can also be increased further. This research aims to investigate whether email phishing during a pandemic has been accelerated and the proposed research highlights that the phishing sensitivity is focused on the protocols utilised in this research. The key purpose is to express a technique or algorithm for the dissection of mailbox information in order to identify it as phishing or to include a genuine email. Machine Learning is a part of Artificial Intelligence (AI), which uses the knowledge mining method to recognise new or current trends (or highlights) of a data set which is then used for characterisation purposes. This study will discuss the advancement and types of phishing attacks. It will examine the Machine Learning techniques and methods which are currently being utilised. The researcher will further analyse a structure on how to avoid phishing as well as recommending methods which can be improved upon for email phishing. Furthermore, the important role of human behaviour is highlighted i.e., working from home during the pandemic.

Y. S. Parmar · H. Jahankhani (✉)
Northumbria University, London, UK
e-mail: Hamid.jahankhani@northumbria.ac.uk

© The Author(s), under exclusive license to Springer Nature Switzerland AG 2021
R. Montasari and H. Jahankhani (eds.), *Artificial Intelligence in Cyber Security: Impact and Implications*, Advanced Sciences and Technologies for Security Applications,
https://doi.org/10.1007/978-3-030-88040-8_3

Keywords Phishing · Data mining · Clustering · Machine learning · Support
vector machine · Naïve Bayes

1 Literature Review

Stealing an individual's identity is perhaps the most mainstream cybercrime activity.
As per Federal Trade Commission in 2015, identity theft was positioned second
with 16% of all client complaints. Stealing personal information online is known
as phishing. PhishTank describes phishing as "a fraudulent attempt, usually made
through email, to steal your personal information".

While the world is centered around the fundamental danger presented by Covid-
19, cybercriminals around the globe without a doubt are ready to exploit the emergency by dispatching an alternate sort of "infection." More employees are working
from home/remotely and organisations may at last face the possibility of working with
almost no faculty on on-site so therefore reducing significant help capacities. Against
this setting, organisations and workers need to take the most extreme consideration to protect themselves from cyber-attacks and keeping the company information
confidential.

In a report (2020), Barracuda Networks stated that phishing emails have accelerated to an unbelievable extent (667% increase) during the pandemic of COVID-19.
The security merchant noticed only 137 occurrences in January, ascending to 1188
in February and 9116 in March. An estimate of 2% of the 468,000 email attacks
worldwide identified by the firm were named COVID-19-themed.

According to Muncaster [18], phishing email attacks of COVID-19 were
categorised into groups as seen in the Table 1.

Just as the standard baits to navigate for more data on the pandemic, a few scam-
mers are professing to sell cures as well as face-covers (masks), while others attempt
to inspire interest in organisations delivering vaccines, or donations to battle the
infection and offer help to casualties.

According to Thomson Reuters [22], $1 million USD was lost to COVID-19
scams since February due to scared shoppers in the UK. Fraudsters were professing
to have the best defensive masks from the virus.

Joe Tidy [23], reports that people are being sent an immense assortment of
messages which mimic specialists, for example, the World Health Organization
(WHO), with an end goal to convince casualties to download programming or give
to false causes, as seen in Fig. 1. Cyber criminals are additionally endeavouring to
exploit government upheld bundles by mimicking public foundations.

Further examination has been carried out on pandemic and working from home as
well as critically analysing machine learning algorithms on phishing emails. The main
objective is to investigate the current techniques and develop a Machine Learning
algorithm which will detect phishing emails.

Table 1 Phishing types in COVID-19

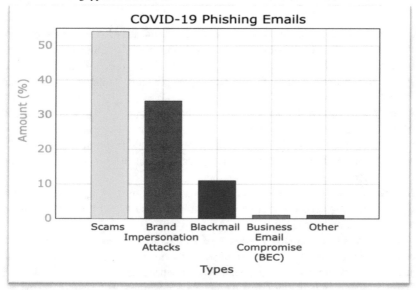

Fig. 1 Phishing email

1.1 Working from Home

Cybercriminals are exploiting employees from organisations who might be less careful while working from home. They are utilising "social engineering" to perpetrate their crimes. Social engineering fraudsters utilise different techniques to complete their assaults such as phishing and spear phishing (targeted at a particular individual) being the most common.

According to AmTrust Financial [6], the thought behind social engineering is to exploit somebody's common inclinations or to inspire a passionate response of "act first, think later." COVID-19 has sadly given adequate freedom to this, as individuals wherever are on edge for data and updates about the infection, excited for communication.

Towards the start of the pandemic, quite possibly the most every now and again used procedures by cybercriminals was the utilisation of Covid disease rate maps with pre-stacked malware [10].

Figure 2 shows how emails were sent to victims to lure them into clicking the links to get most recent updates for example. This made them eagerly open the document without thinking about the dangers. When opened, the record would download malicious malware, which can infect the device for example a computer and cybercriminals can then gain sensitive information. Messages were conveyed promising beneficiaries they could get their checks quicker. Clients were approached to enlist their email address or other individual data Brewster [10].

Cialdini (2007), explains that there are seven principles of influence which also translate to phishing:

Reciprocation—An email promising to offer admittance to "confidential" data if the attachment is downloaded from the email (utilising phishing essentially).

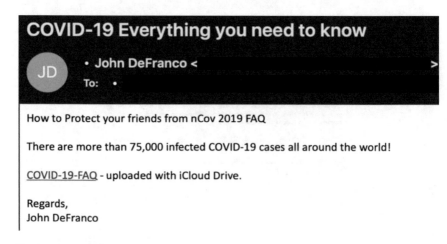

Fig. 2 COVID-19 Phishing email

Scarcity—Phishing messages that pressure that a specific advantage is available just if a move is made inside a brief timeframe.

Authority—For example, an email from the CEO asking the finance department to transfer £100 k to an unknown account.

Consistency—Phishing messages that seem as though official interchanges abuse this reality, trusting the beneficiary disregards the unordinary demand that is made in the email.

Consensus—People follow people. If a phishing email specifies that 600 of 700 people have downloaded this software (example) "click here to download". This will make individuals want to click on the link.

Liking—If individuals like you then they say "Yes". This is a technique used by phishers as they try to satisfy new workers.

Unity—A phishing email probably sent by somebody who has similar interests as the recipient, data that can without much of a stretch be sourced through social media, normally has a good success rate.

1.2 Advancement of Phishing Attacks

The first main case of this method (phishing) was accounted for in 1995 when attackers utilised phishing to persuade victims to share their AOL account information.

Phishing however has evolved and has advanced massively impacting millions around the world by costing billions of dollars to organisations.

Other attacks apart from email phishing have been conducted in the past. In the time of the 2000s, instant messenger (IM) was being used all around the world to communicate. Cyber criminals in May 2006 sent numerous spam messages via IM accounts. Recipients thought that messages came from someone they knew and they clicked the connection which would request the users email address and password. Once the information was entered, attackers were able to access all their records and will start to convey a similar process with the user's contacts to build trust. This was a computerised interaction that affected a huge number of IM clients in 2006.

By figuring out how to perceive a phishing attack before it starts, employees can settle on more viable choices for the security of the organisation. Phishing has evolved into specific categories as cybercriminals strategize and take phishing to another level.

Many strategies for the identification of phishing emails have been developed, ranging from the communication-oriented approaches such as authentication protocols and blacklisting to content-based filtering [20]. The strategies of blacklists and white lists have not proved to be effective enough in various regions, and are not widely used.

In the meantime, phishing filters based on content have been used extensively and shown to be highly effective. With that in mind, research focuses on content-related mechanisms and the development of computer and data mining methods based on the e-mail headers and body Kumaraguru [16].

In a report in 2009, the majority of resources against phishing didn't start to block phishing websites until many hours/days after these phishing emails were sent to consumers. This indicates that the security tools presently in use do not fully detect these emails and websites 100% [16]. This chapter introduces numerous algorithms to detect and forecast Phishing Emails.

In 2007, Abu-Nimeh discussed various machine learning techniques and compared the accuracy of their predictions using phishing techniques and 2889 genuine emails. The research contained the following techniques i.e., Logistic Regression (LR), Classification and Regression Trees (CART), Bayesian Additive Regression Trees (BART), Vector Support Machinery (SVM), Random Forests (RF), and Neural Networks (NNet) that were also checked with 43 functions. The findings show that the error rate for RF is above all the other categories, followed by CART, LR, BART, SVM, and NNet, respectively, with an error rate of 07.72 percent, provided that the legal and phishing emails have the same value. The LR at a 4.89 percent average was the highest result for false positives, followed by BART, NNet, CART and SVM, while the RF was at a rate of 08.29 at its worst false positive rate [1].

In addition, Bergholz proposed to distinguish e-mails in which two new features were generated, both the adaptive Dynamic Markov Chain (DMC) approach and the latent Class-Topic Model (CLTOM) were proposed by Bergholz. In contrast to the standard edition, the adaptive version of DMC accomplished in delivering the same consistency with using two thirds less storage. The adaptive edition of the CLTOM has demonstrated a better performance than the regular LDA as the first introduces class-specific data into the subject model [9].

In order to sort into the phishing or non-phishing groups, Toolan created a new C5.0 method choosing 5 attributes. The means information included 8,000 emails, half were phished and the other half were genuine. This method beats the output in terms of increased retrieval performance of any other human classifier or collector [24].

Dr. Ma used an application with a sequence of spelling features to dynamically cluster phishing mails and delete redundant and irrelevant features. The combination of this strategy and the collection of features has obtained highly productive performance. Ma used the global K-mean model with a minor adjustment and produced the values of the target function over many tolerance values in selected subsets of characteristics. The objective function values helped to define the relevant clusters depending on the distribution of these values [17].

Jameel and George have used a neural feedback method to classify the phishing email by deleting features from the email and HTML bodies. The algorithm they proposed was checked with 18 characteristics using five hidden neurons. A training software takes 173.55 ms for this algorithm before it is implemented. The time to validate an email is 0.00069 s. With the increase in the number of neurons, the time spent will increase while it is still considered minimal. The algorithm displayed an outstanding accuracy of 98.72% in terms of performance and a study rate of 0.01 [13].

Dr. Wu worked on spoofing emails and Microsoft Outlook facilities by designing a sender authentication protocol (SAP). By checking demanded sender1 with archived addresses, it confirms the legitimacy of the sender. The OutlookTM improved has an add-in that checks its performance while the same user-friendly interface has stayed on the initial version and the SAP add-in is automatically started as long as the OutlookTM runs [25].

Zhang's goal was to test the performance of the phishing email identification cross-validation strategy. He has used multi-layer neural network (NN) systems with multiple hidden unit quantities and activation functions to show that NNs with an approximate hidden unit number can provide reasonably reliable and effective performance. It should be remembered that, even with little experience, he proved these findings while evaluating the functions set to produce better results [26].

To test the accuracy of the email classifier, Kumar has used TANAGRA data mining on a sampled spam data set, using multiple algorithms. In the end, the Fisher spam filters and Rnd filtering functions received higher scores. Pursuing fisheries filtering that has achieved more than 99 percent spam identification accuracy, an algorithm was used on the required features for the Rnd tree classification [15].

Al Momani identified a new method in 2013, which showed excellent performance with regard to real positive, true negative, sensitivity, precision, F-size and overall accuracy in comparison to other methods. The framework demonstrated reliability in the on-line prediction of these messages and the long life with memory usage on footprint. The model Al Momani designed to forecast and identify unknown phishing emails on zeros day, named Phishing Dynamic Evolving Neural Fuzzy Framework (PDENF) [5].

Pandey identified phishing emails using different techniques such as; Multilayer Perceptron (MLP), Decision Trees (DT), Supporting Vector Machine (SVM), and Logistic Regression (LR). This mix was designed to parallel the use of text and data mining for detection, where 23 keywords were omitted from the email box and the data samples were included, as well as a total of 2500 phishing and non-phishing e-mails were examined.

The 12 most powerful functions in forecasting phishing emails with precision and minimum number are used to complete a predictive functional range. The research compared method effects with and without the selection of functions. As a result, no impact on the classification systems, and the detection mechanism was shown by the choice of features. This was rational because, with or without the collection of functions, the GP and DT do not vary statistically. The DT however follows the 'if–then' law, which functions as a system for early warning experts it is thus preferable and used commonly.

With respect to the classification of websites, Khonji checked the updated approach for avoiding the usage and reliability of phishing emails. The technology proposed previously was based on a lexical analysis of the website URLs which improved the filter accuracy by 97%. The analyses of Lexical URL demonstrated greater specificity in the classification of anti-phishing [14].

Al Momani and others have found a further mechanism that also relies on an "evolving connecting system" to locate unknown zero-day phishing emails. The new system was called PDENFF, the phishing dynamic evolving neural fuzzy framework, and is assisted by an offline training feature to accomplish the intended objective, adopting a hybrid learning method (supervised / non-supervised). With this method, the identification of zero-day phishing messages has been increased from 3 to 13%. In addition, rules, categories or functionality were used to improve the learning experience using ECOS to differentiate the framework from legitimate emails.

In order to detect the unknown zero-day phishing emails by managing all related function vectors for prediction purposes, Altaher relied on Adoptive Evolving Fuzzy Neural Network (EFuNN) for making Phishing Evolving Neural Fuzzy Framework (PENFF). The PENFF solution, therefore, depends on the parallels between the roles used in the corporate and the URL of the message.

A framework for classifying phishing emails using a forest learning mechanism was developed later in 2014 by Akinyelu. This mechanism has been checked on data containing around 2000 advanced phishing emails, and it has also been able to distinguish high efficiency (99.7%) phishing emails at low false negatives and false positives. Akinyelu's algorithm is thus more effective, requiring fewer characteristics to detect phishing and more precise performance.

Nizamani et al. [19] has introduced a fraudulent detection model that uses an innovative range of features where malicious email detection rates have been compared to the individual categories. In addition to the various feature sets the analysis used many classification methods and algorithms, such as SVM, NB, J48, and CCM. A 96% precision was reached and the results revealed that the exactness was influenced instead of the classifiers by the form of functionality.

In 2015, Kathirvalavakumar and others suggested a neural network with multilayer for emails to be detected. His proposed network depends on a feedback algorithm to collect distinguished knowledge and characteristics from the e-mail, which uses the technique of weight reduction. This pruning method leads to the minimal estimate expected to identify emails to Phishing by decreasing the number of features across the algorithm. In terms of false-positive outcomes and false-negative results, the network has obtained strong results. As this framework was tested on 2007 results, it needs to acknowledge the latest features in the algorithm integrating them in the training input domain in order to be useful in using this network for current knowledge.

1.3 Traditional Methods

Modern identification methods fall into two groups, security at the network level and protection against encryption. The first type of network security involves blacklist filters and white-list filters that restrict the entry of suspected IP addresses and domains into the network by preventing physical action [21]. There are also pattern matching filters and controlled filters that rely on set detection regulations that are individually submitted and modified.

1.3.1 Blacklist Filter

The blacklist filter technology offers network security by the identification of emails collected based on the address of the sender, the IP address, or the DNS address. This information is retrieved from the header of the email and compared to the predetermined set, and the email would be refused if one of these data falls in line with the list. This strategy then filters phishing emails to provide network protection. This filter is being supplied and introduced by Internet Server Providers (ISP) [20].

1.3.2 Whitelist Filter

The filtering of the whitelists offers network security too, although this strategy contrasts the data of email with a predefined list that includes static IP addresses of valid domains and IPs, unlike blacklists. In that respect, the user's mailbox can only be opened via e-mails with data corresponding to the list [11]. If they belong to legally binding users or corporations who have decided to add their addresses to this registry, email addresses, and IP addresses are included in the whitelist. Emails with the data to be matched to this list are only regarded as valid depending on this filter, while the phishing and entry to the network for which this filter is often called legitimate email classificatory are considered for other emails.

1.3.3 Email Verifications

Email verification is a device authentication process involving sender and recipient verification. The e-mail is accredited and eligible to be legitimately passed to the receiver's mailbox when the sender accepts the post. Otherwise, the email would be viewed as phishing such that access to an inbox is not enabled. The filter has its benefits and drawbacks. While this filtering mechanism has proved successful in totally identifying phishing emails (100%), it also takes a long time to respond to the user before the message is retrieved and there is a possibility that the e-mail will be missed if traffic across the network is not noticed or the same challenge is not identified in the verifying process [3].

1.3.4 Password Filter

Filters with passwords offer user security protection. The filter helps you to accept any email on the subject line, email address, header or email address, provided only the filter has been able to detect the password you have determined. Therefore, the email would be refused if the filer cannot recognize a password or identify an incorrect password. This password is not created by default, so the first-time users of this filter initiate a chat to set up and turn on a password and then be listed as legit by the filter [21]. This method of the filter also has its flaw in that some legal emails may be lost if the password is not remembered, plus the procedure takes time.

1.4 Machine Learning Algorithms to Detect Phishing

This approach refers to automatic classification systems that focus on computer and information management. These classifiers run alongside the server and process received emails in phishing or legitimate by evaluating the various features if the email is header and body [1].

1.4.1 Logistic Regression

Due to its easily understood and functional effects, logical regression is a commonly used technique. This model is functional in the prediction of binary data (0/1 response) since it uses a simplified linear model that uses statistical data. In terms of its simplicity, this approach has three weaknesses i.e., first, before it is implemented, it needs more mathematical assumptions. Secondly, it is more functional for linear variables than those with a complex relationship. Finally, the forecast rate is reliable and sensitive to data integrity [1].

1.4.2 Decision Trees Filter (DT)

Decision Trees Filter is a visual grouping model composed of nodes and arrows and the base node is considered as the Root. The decision tree ends with the terminator node. The tree may be composed of one or more classifier steps, and internal nodes are bounded by root and terminating nodes. During the evaluation of the aim, different algorithms were proposed for the development of decisions including the ID3 model that calculates attrition information as a heuristic function. In this way, the Decision Tree creates sub-trees, with each node in the tree having its parent node and each one leading to a child node as well. The final node represents the solution of the proposed problem and the tree ends with the terminating node.

1.4.3 Support Vector Machine (SVM)

Medical diagnosis, text classification, classification of images, bio-sequences analysis, and other areas of study, SVM is commonly applied by researchers. This methodology uses numbers, square equations, and sets rules to separate data into 2 groups.

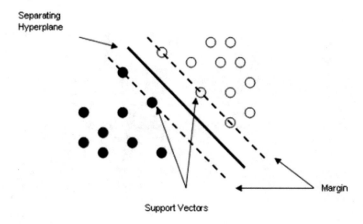

Fig. 3 SVM Hyperplane [21]

A separate hyper-plane for improving the space for the margin base of the kernel functions and collecting and storing data into the vector would establish, and the binary data classification would find the right solution to the problem and find the required classification.

The support vectors are seen on the boundaries in Fig. 3. The hyperplane division lies in the middle of the margin, enhancing the separating margin [21].

The most important monitoring methods like the Cloudmark, Netcraft, FirePhish, the eBay Account Guards, and the IE Phishing Filter are summarized in Table 2. The major drawbacks of commonly used common tools are also mentioned.

1.5 Current Prevention Techniques

Gansterer has introduced a screening framework which classifies emails obtained into 3 groups, i.e., legal, malicious and phishing, using the features from these emails that were newly created. The scheme contains multiple classifiers for the categorization of messages received. Among these three classes, a classification accuracy of 97% has been achieved that is better considered than a series of two binary graders to deal with the ternary classification problem [12].

Basnet explores an approach for identification that incorporates readily obtained functionality from the text of the email without focusing on heuristic functionality. Phishers created images from the text of the post, which passes the phishing filter only with graphic details [8].

Alguliev has created a new method for spam clustering and problem-solving. The suggested algorithm implements the technique of enhancing the similarities of the texts in the clusters and determines the optimal solution by the opposing algorithm. But the continued help of chromosomes limits certain algorithms, which decreases

Table 2 Phishing detection tools

Tool	Type	Description	Advantage	Disadvantage
Snort	Network level	Heuristic tool	Good at detecting level attacks	Rules require manual adjustments. Does not look at content
Spam Assassin	Server Side Filter	Heuristic engine uses specific features	Good at detecting email header spoofing	High false positives
PILFER	Server Side Filter	Utilize 10 features	Better performance than spam assassin	Did not use content from the body of the email. Used with short lived phish domains.
Spoof Guard	Client Side Tool	Plug-in to a browser	Warns user if link points to phishing site	Users do not pay attention to warnings. Not all email clients are browser based.
Calling ID, Cloud Mark, Netcraft, and Fire Phish	Client Side Tool	Utilizes blacklist of domains	Good for domains that employ domain level authentication	Phish domains are short lived. Does not look at email content.
eBay Account Guard	Client Side Tool	Utilizes blacklist of eBay URLs	Protects eBay users.	Specific website tool.
IE Phishing Filter	Client Side Tool	Records specific user website visiting patterns.	Adapts to user website visit pattern	Works only on internet explorer.
Catching Phish	Client Side Tool	Detects fake website based on rendered images	Browser independent. Good results on small data sets.	Processing time is high. Susceptible to screen resolution

the mechanism for optimization of attempting to solve problems. A penalty feature is then used to speed up the process of convergence and avoid infallible chromosomes. The resulting grouping is then examined in detail in order to complete class details and an informative portrait is created by documents to help explain the groups and scam messages. In addition to the study of the sources of spam communications, this antispam method can help to anticipate threats on the targeting of spammers [4].

Azad has based its attention on the consistency testing of many proposed methods, such as Naive Bayes, logistic regression, and SVM. In general, high findings were found suggesting an accuracy rate of 95% for SVM with a linear kernel and Bayes over the other classifiers, as only 10 and 2, 66% of phishing emails were incomplete. The SVM showed similar outcomes with fewer features as compared with Naive Bay and logistical regression. Meanwhile, a linear SVM was checked and additional characteristics omitted to result in better identification rates, as 5.86% of Phishing emails were misclassified, which indicates that additional characteristics increase the consistency of the results. The study showed, in conclusion, that linear SVM helps to spot phishing emails before they even enter the user's inbox [7].

1.6   Framework of Preventing Phishing

Preventing phishing refers to using a set of tools and methods which will help to distinguish and end phishing attacks before they are made. This incorporates user education to spread awareness of phishing, installing anti-phishing solutions, programs and presenting various other phishing safety efforts that are focused on proactive phishing assurance while giving mitigation procedures to attacks that do figure out how to penetrate security.

User education is key. Training improves one's information, abilities and builds up the character and mentality. Educating people on how phishing emails look like and what type of common things to look for in the emails will help reduce phishing attacks as user's will be more aware of what to identify such as:

Poor spelling/grammar in email—It is likely that a phishing email will have basic errors such as spelling and grammar. This is common as some attackers use Google Translate to translate the message into English for example, as their first language may be different. Moreover, it is highly unlikely that official organisations will send out poorly written emails.

URL (Link)—Most phishing emails provide website links to have the user click onto a malicious website. Some website links have been shortened (URL) where attackers would hope that the user thinks it's a real link and other links are long. When you hover over a link to check it seems different and has symbols inside it. Asking for personal information—Some emails ask users to verify their details acting as a legitimate business such as PayPal. The email will ask users to verify banking information etc.

Strange rewards—Some emails normally overwhelm users by saying "Congratulations", you've won a car or lottery. They will ask the user to provide banking information or sensitive data. The best thing to do is ignore these as they are too good to be true (ZDNet).

Urgent deadlines—Some hackers will send deadline emails such as "your policy is expiring soon" so the user will click the link and input their personal information which is valuable to the hacker.

There are many more aspects to identifying phishing emails. So, user training is essential and organisations should teach employees these details. Another way of protecting and having a strong gateway up-against phishing is Multi-factor Authentication. This is a way of verifying the account of a user by not just entering a password to login but also entering a text code (which comes via phone) or answering a security question to login. This creates a stronger account and becomes harder for attackers to attack. Microsoft states that multi-factor authentication blocks 99.9% of account hacking.

Using specialised anti-phishing software also helps to prevent phishing. Using these software's and tools helps to handle zero-day weaknesses, recognising and protecting against malware, identifying spear phishing messages and acting as a firewall before they get to the users. This will automatically prevent phishing emails from reaching the users. Use HTTPS which are secure websites especially when dealing with sensitive information. The use of public networks should be limited (certainly for banking and personal information).

2 Software Methodology

2.1 Introduction

This section presents the work which is proposed for phishing detection. It will discuss the software methodology chosen and what are the advantages and limitations of it. Furthermore, the machine learning algorithm is critically analysed to detect email phishing. Machine Learning has been used to run automated malicious filters to efficiently use emails. The aim is to find trends through the classification algorithm of machine learning so that it can recognise emails such as malicious or not malicious.

2.2 Software Development Methodologies

Projects which are successful are always managed well. To deal with a task proficiently, the best software development methodology is chosen which will work best for the project. Every methodology has their advantages and disadvantages. They are utilised for different purposes according to the need of the project. The two most used methodologies and their strategies are as follows.

2.2.1 Agile Development Methodology

Agile techniques break tasks into little increments and no future planning is conducted. Each phase of development is returned to throughout the lifecycle of the project continually in sprints (iterations). Sprints are time frames lasting around 1 to 4 weeks. This "assess and-adjust" approach fundamentally decreases both improvement expenses and time to advertise.

Every sprint includes working through the life cycle of the software as seen in Fig. 3. This will start by planning and designing. Once the design is complete the moving onto the development phase and then testing of the development. Then the system is released and feedback is acquired to improve. The sprints make the release of the system with minimal errors.

The main advantages of agile methodology is that the software can be produced in sprints (iterations). This allows efficiency as errors can be detected and fixed in early stages. Frequent improvements are continually made.

maliciousness	emails
0	not_malicious	Go until jurong point, crazy.. Available only in bugis n great world la e buffet... Cine there got amore wat...
1	not_malicious	Ok lar... Joking wif u oni...
2	malicious	Free entry in 2 a wkly comp to win FA Cup final tkts 21st May 2005. Text FA to 87121 to receive entry question(std txt rate)T&C's apply 08452810075over18's
3	not_malicious	U dun say so early hor... U c already then say...
4	not_malicious	Nah I don't think he goes to usf, he lives around here though
5	malicious	FreeMsg Hey there darling it's been 3 week's now and no word back! I'd like some fun you up for it still? Tb ok! XxX std chgs to send, £1.50 to rcv
6	not_malicious	Even my brother is not like to speak with me. They treat me like aids patent.
7	not_malicious	As per your request 'Melle Melle (Oru Minnaminunginte Nurungu Vettam)' has been set as your callertune for all Callers. Press *9 to copy your friends Callertune
8	malicious	WINNER!! As a valued network customer you have been selected to receivea £900 prize reward! To claim call 09061701461. Claim code KL341. Valid 12 hours only.
9	malicious	Had your mobile 11 months or more? U R entitled to Update to the latest colour mobiles with camera for Free! Call The Mobile Update Co FREE on 08002986030
10	not_malicious	I'm gonna be home soon and i don't want to talk about this stuff anymore tonight, k? I've cried enough today.
11	malicious	SIX chances to win CASH! From 100 to 20,000 pounds txt> CSH11 and send to 87575. Cost 150p/day, 6days, 16+ TsandCs apply Reply HL 4 info
12	malicious	URGENT! You have won a 1 week FREE membership in our £100,000 Prize Jackpot! Txt the word: CLAIM to No: 81010 T&C www.dbuk.net LCCLTD POBOX 4403LDNW1A7RW18
13	not_malicious	I've been searching for the right words to thank you for this breather. I promise i wont take your help for granted and will fulfil my promise. You have been wonderful and a blessing at all times.
14	not_malicious	I HAVE A DATE ON SUNDAY WITH WILL!!
15	malicious	XXXMobileMovieClub: To use your credit, click the WAP link in the next txt message or click here>> http://wap. xxxmobilemovieclub.com?n=QJKGIGHJJGCBL
16	not_malicious	Oh k...i'm watching here:)
17	not_malicious	Eh u remember how 2 spell his name... Yes i did. He v naughty make until i v wet.
18	not_malicious	Fine if thatå the way u feel. Thatå the way its gota b
19	malicious	England v Macedonia - dont miss the goals/team news. Txt ur national team to 87077 eg ENGLAND to 87077 Try:WALES, SCOTLAND 4txt/£1.20 POBOXox36504W45WQ 16+
20	not_malicious	Is that seriously how you spell his name?
21	not_malicious	Iå'm going to try for 2 months ha ha only joking
22	not_malicious	So ÅÅ, pay first lar... Then when is da stock comin...
23	not_malicious	Aft i finish my lunch then i go str down lor. Ard 3 smth lor. U finish ur lunch already?
24	not_malicious	Ffffffffff. Alright no way I can meet up with you sooner?
25	not_malicious	Just forced myself to eat a slice. I'm really not hungry tho. This sucks. Mark is getting worried. He knows I'm sick when I turn down pizza. Lol
26	not_malicious	Lol your always so convincing.
27	not_malicious	Did you catch the bus ? Are you frying an egg ? Did you make a tea? Are you eating your mom's left over dinner ? Do you feel my Love ?
28	not_malicious	I'm back & we're packing the car now, I'll let you know if there's room
29	not_malicious	Ahhh. Work. I vaguely remember that! What does it feel like? Lol
30	not_malicious	Wait that's still not all that clear, were you not sure about me being sarcastic or that that's why x doesn't want to live with us
31	not_malicious	Yeah he got in at 2 and was v apologetic. n had fallen out and she was actin like spoilt child and he got caught up in that. Till 2! But we won't go there! Not doing too badly cheers. You?
32	not_malicious	K tell me anything about you.
33	not_malicious	For fear of fainting with the of all that housework you just did? Quick have a cuppa
34	malicious	Thanks for your subscription to Ringtone UK your mobile will be charged £5/month Please confirm by replying YES or NO. If you reply NO you will not be charged

Fig. 4 Training data set

The disadvantage of agile is that it has limited documentation as it lacks real time communication because developers are mostly focused on iterations.

2.2.2 Waterfall Development Methodology

Waterfall method is an inflexible direct model that comprises consecutive stages which directly concentrate on the objective. Figure 4 shows the stages of the methodology. Each stage should be 100% finished before the following stage can begin. There's normally no cycle for returning to adjust the previous.

The advantages are that this method is very easy to understand and as it is simple, projects can be managed easily. It sets out clear stages to follow.

The disadvantage is very rigid to follow. It gives a slower process and has too much control as each stage has to be completely finished before moving.

2.2.3 Selection of Methodology

Due to the nature of this project which is creating an algorithm to detect email phishing, agile methodology has been selected. The reason for this is because the agile method gives an option to work in iterations which means each stage will be improved continuously to fix errors. This assures the quality of the development. For this project continuous improvements need to be made after testing the algorithm with the dataset. There are limitations of agile which have been considered by the author and have applied ways to overcome the limitations. The machine learning algorithm documentation is provided and all results have been outlined.

2.3   Guide on Creation of the Machine Learning Algorithm

Various ML algorithms were trained and worked on to detect email phishing. The tool in which the algorithm was created is called Jupyter Notebook. Python was used as the core language to program the algorithm.

To begin a product backlog was created listing the deliverables which will be implemented during the development phase. This helped with the prioritisation.

2.3.1   Learning Data (First Sprint)

The proposed approach to classifying phishing e-mail uses a model of Information Exploration and Data integration to construct an intelligent email classification capable of classifying a new email message as valid or spam; the proposed model is designed to define and extract useful features from training email data by using the steps of the support vector machine. In order to increase similitude between email messaging with the same semantic word sense, the proposed email classification model uses linguistic analysis strategies and ontology. In addition, the phishing words in a single email are weighted in such a way that email phishing terms weighing help distinguish phishing from valid emails.

A user may be able to screen a massive ordered channel and detect a pattern, but computers can analyse even larger datasets with machine learning as they can recognise connective patterns much better than any human-created channel feature can.

The data used by the various email users for this project was taken. The dataset which was collected included almost all manner of malicious and non-malicious emails, as well as the Google Mail spam & post archive (GMAIL). The dataset comprises a randomly chosen group of addresses, whether MALICIOUS or NOT MALICIOUS in plain text format. Training data was used to create the MALICIOUS and NOT-MALICIOUS email detection model. A total of 5572 e-mails have been obtained, which combine malicious and non-malicious mail.

Learning data set was imported into a excel sheet into 2 columns one showing the maliciousness and the other column shows the email, as seen in Fig. 4.

The data was then imported to Jupyter to create the algorithm and train the dataset as seen in Fig. 5.

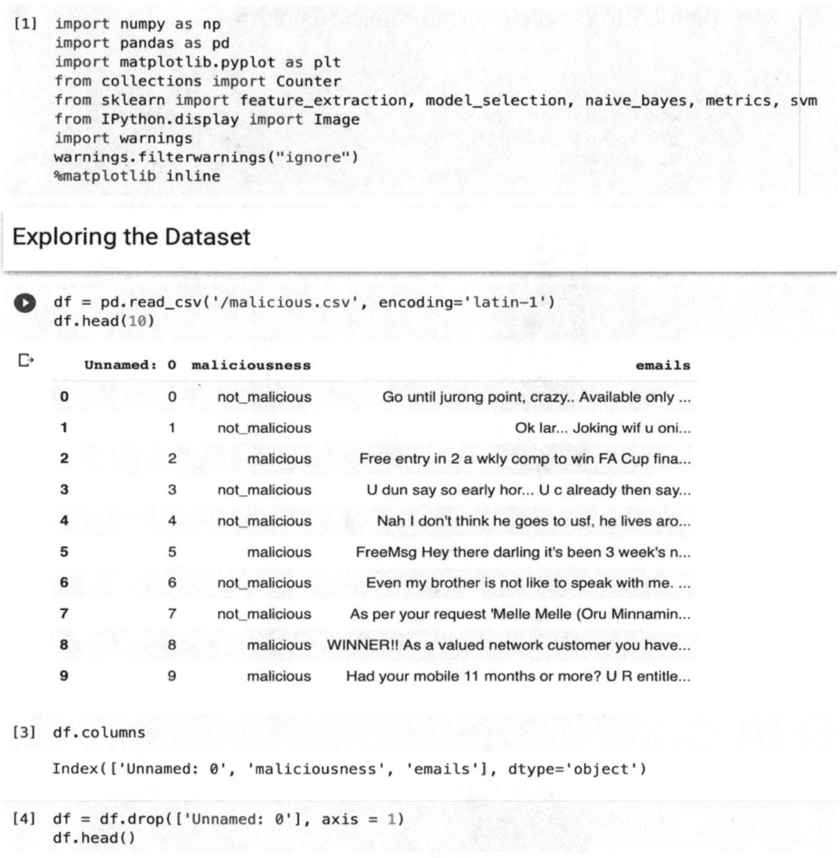

```
[1] import numpy as np
    import pandas as pd
    import matplotlib.pyplot as plt
    from collections import Counter
    from sklearn import feature_extraction, model_selection, naive_bayes, metrics, svm
    from IPython.display import Image
    import warnings
    warnings.filterwarnings("ignore")
    %matplotlib inline
```

Exploring the Dataset

```
df = pd.read_csv('/malicious.csv', encoding='latin-1')
df.head(10)
```

	Unnamed: 0	maliciousness	emails
0	0	not_malicious	Go until jurong point, crazy.. Available only ...
1	1	not_malicious	Ok lar... Joking wif u oni...
2	2	malicious	Free entry in 2 a wkly comp to win FA Cup fina...
3	3	not_malicious	U dun say so early hor... U c already then say...
4	4	not_malicious	Nah I don't think he goes to usf, he lives aro...
5	5	malicious	FreeMsg Hey there darling it's been 3 week's n...
6	6	not_malicious	Even my brother is not like to speak with me. ...
7	7	not_malicious	As per your request 'Melle Melle (Oru Minnamin...
8	8	malicious	WINNER!! As a valued network customer you have...
9	9	malicious	Had your mobile 11 months or more? U R entitle...

```
[3] df.columns

    Index(['Unnamed: 0', 'maliciousness', 'emails'], dtype='object')
```

```
[4] df = df.drop(['Unnamed: 0'], axis = 1)
    df.head()
```

Fig. 5 Data imported and used in python

2.3.2 Categorisation of the Data (Second Sprint)

The learning data was further categorised into the two groups to show the distribution of malicious and non-malicious emails as seen in Fig. 6. In addition, emails were visualised into a bar chart to show the number of emails being used of the data set as seen in Fig. 6 below.

2.3.3 Pre-processing (Third Sprint)

The emails in the learning data are in plain text format so therefore a conversion from the plain text into features has been conducted to represent the emails. This is because one these features are in place an algorithm can be used. A number of pre-processing steps are first performed.

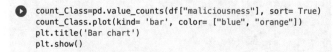

```
count_Class=pd.value_counts(df["maliciousness"], sort= True)
count_Class.plot(kind= 'bar', color= ["blue", "orange"])
plt.title('Bar chart')
plt.show()
```

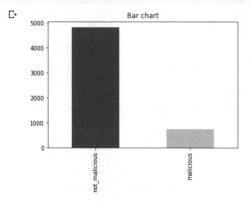

```
[6]  count_Class.plot(kind = 'pie',   autopct='%1.0f%%')
     plt.title('Pie chart')
     plt.ylabel('')
     plt.show()
```

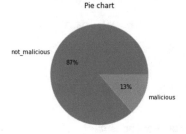

Fig. 6 Pie & bar chart of data

The first step was to do Text Analytics to find out the frequencies of words in the malicious and not-malicious emails. The words of the emails will be model features. The 'Counter' function of Python has been used. Below is the visualisation of the most frequent words in the not-malicious emails (Fig. 7).

Words such as 'to, you, I, the, a, and, i, in, u, is, my, me, of, for, that, it, your, on, have, at' are the most frequent words in the not-malicious emails (Fig. 8).

Words such as 'to, a, your, call, or, the, 2, for, you, is, call, on, have, and, from, Ur, with, &, 4, of' are the most frequent words in the malicious emails.

Majority of frequent words in both classes are stop words such as 'to', 'a', 'or' and so on. These are common stop words in a language so therefore these words have to be removed before creating features model.

```
[7]  count1 = Counter(" ".join(df[df['maliciousness']=='not_malicious']["emails"]).split()).most_common(20)
     df1 = pd.DataFrame.from_dict(count1)
     df1 = df1.rename(columns={0: "words in not-malicious", 1 : "count"})

     count2 = Counter(" ".join(df[df['maliciousness']=='malicious']["emails"]).split()).most_common(20)
     df2 = pd.DataFrame.from_dict(count2)
     df2 = df2.rename(columns={0: "words in malicious", 1 : "count_"})
```

```
df1.plot.bar(legend = False)
y_pos = np.arange(len(df1["words in not-malicious"]))
plt.xticks(y_pos, df1["words in not-malicious"])
plt.title('More frequent words in not-malicious emails')
plt.xlabel('words')
plt.ylabel('number')
plt.show()
```

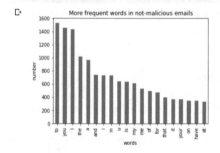

```
[9]  df2.plot.bar(legend = False, color = 'orange')
     y_pos = np.arange(len(df2["words in malicious"]))
     plt.xticks(y_pos, df2["words in malicious"])
     plt.title('More frequent words in malicious emails')
     plt.xlabel('words')
     plt.ylabel('number')
     plt.show()
```

Fig. 7 Not Malicious common words

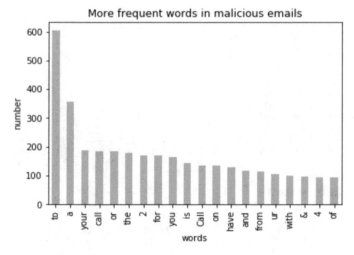

Fig. 8 Malicious common words

```
[10] f = feature_extraction.text.CountVectorizer(stop_words = 'english')
     X = f.fit_transform(df["emails"])
     np.shape(X)

     (5572, 8397)
```

Fig. 9 Stop words removal

The next step is Feature Engineering. Text pre-processing, tokenising, and filtering of stop words are included in a high-level component that is able to build a dictionary of features and transform documents to feature vectors. To improve the analytics the stop words have been removed as shown in Fig. 9.

2.3.4 Predictive Analysis (Fourth Sprint)

The main Machine Learning algorithms which are used are Support Vector Machine (SVM) and Multinominal Naïve Bayes.

Firstly, different bayes model were trained (regularisation changed to parameter a) as seen in Fig. 10.

Figure 11 shows that the models were trained and tested to check their precision and accuracy. Before building the model, the dataset was split into train & test sets. 33% for the testing set & 77% for the training set. After splitting, the distribution of emails in both training & testing test is given below:

Training Set: 3733 emails.

```
[12] list_alpha = np.arange(1/100000, 20, 0.11)
     score_train = np.zeros(len(list_alpha))
     score_test = np.zeros(len(list_alpha))
     recall_test = np.zeros(len(list_alpha))
     precision_test= np.zeros(len(list_alpha))
     count = 0
     for alpha in list_alpha:
         bayes = naive_bayes.MultinomialNB(alpha=alpha)
         bayes.fit(X_train, y_train)
         score_train[count] = bayes.score(X_train, y_train)
         score_test[count]= bayes.score(X_test, y_test)
         recall_test[count] = metrics.recall_score(y_test, bayes.predict(X_test))
         precision_test[count] = metrics.precision_score(y_test, bayes.predict(X_test))
         count = count + 1
```

Fig. 10 Naïve Bayes classifier

```
[13] matrix = np.matrix(np.c_[list_alpha, score_train, score_test, recall_test, precision_test])
     models = pd.DataFrame(data = matrix, columns =
                 ['alpha', 'Train Accuracy', 'Test Accuracy', 'Test Recall', 'Test Precision'])
     models.head(n=10)
```

Fig. 11 Metrics of learning model

Testing Set: 1839 emails.

2.4 Ethical Viewpoint

Putting machine learning into context of an ethical approach certain key factors must be considered. The algorithm created to detect email phishing can be questioned as many researchers question the fact that an algorithm is making a decision. However, an algorithm in this research has been developed to protect against attackers so that personal information is not lost.

The decisions of an algorithm are simply a mathematical judgement to specific emails. The way this algorithm is implemented is not an ethical certain as it is not collecting any private data of users. The algorithm is a tool to be in place so that it can have positive impacts.

2.5 Legal Viewpoint

Privacy is one of the main key elements to consider. The volume and relativity of information being collected will keep security at the front line as perhaps the main legitimate issues that ML users will confront going ahead. ML frameworks utilise tremendous measures of information; consequently, as more information is utilised more inquiries are raised. For example, who owns the data? Is the data being shared with anyone? Is the data sold?

In this research and development of the machine learning algorithm, it complies with the GDPR law and standards. No private information has been acquired or shared at any point.

2.6 Social Viewpoint

Security and privacy are amongst the social issues of ML. People will question about if they can even trust an algorithm. They will be the first to blame in case the algorithm has failed to detect a phishing email. The "blame" is part of society and if anything goes wrong then questions are arising straight away. Society always considers the privacy an issue even though people do not make themselves more aware to issues like phishing for example.

Technology advancements such as phones and the internet have ruined the social interaction and social integration is still on the decline. A machine learning algorithm will help to reduce the number of attacks.

3   Results and Critical Discussions

3.1   Results & Analysis

The flow diagram (Fig. 12) shows how the classifier predicted the results:

Firstly, Multinomial Naïve Bayes classifier was implemented to try and determine how accurate it is. Different Bayes models were trained changing the regularization parameter α. The accuracy, recall, and precision of the model with the test set was then evaluated to show the first 10 Multinomial Naïve Bayes models & their metrics (Table 3).

Next the models test precision was checked (≈ 1). This is because the model which has the Test Precision ≈ 1 and has greater test accuracy then that model will be considered as the best model. Figure 13 shows the top 5 models which have test precision ≈ 1.

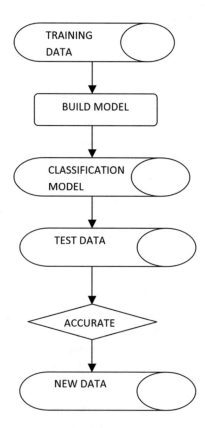

Fig. 12 Flow chart

Table 3 Naïve Bayes classifier

	alpha	Train Accuracy	Test Accuracy	Test Recall	Test Precision
0	0.00001	0.998661	0.974443	0.920635	0.895753
1	0.11001	0.997857	0.976074	0.936508	0.893939
2	0.22001	0.997857	0.977705	0.936508	0.904215
3	0.33001	0.997857	0.977162	0.936508	0.900763
4	0.44001	0.997053	0.977162	0.936508	0.900763
5	0.55001	0.996518	0.976618	0.936508	0.897338
6	0.66001	0.996518	0.976074	0.932540	0.896947
7	0.77001	0.996518	0.976074	0.924603	0.903101
8	0.88001	0.995982	0.976074	0.924603	0.903101
9	0.99001	0.995982	0.976074	0.920635	0.906250

	alpha	Train Accuracy	Test Accuracy	Test Recall	Test Precision
141	15.51001	0.979641	0.969549	0.777778	1.0
142	15.62001	0.979641	0.969549	0.777778	1.0
143	15.73001	0.979641	0.969549	0.777778	1.0
144	15.84001	0.979641	0.969549	0.777778	1.0
145	15.95001	0.979373	0.969549	0.777778	1.0

Fig. 13 Test precision

Between these models with the highest possible precision, a selection of the most accurate one will be made. The following model will be considered as the best naïve Bayes model if it gives the highest accuracy (Fig. 14).

The accuracy of the best Naïve Bayes Classifier is 96% & it classifies 77% of malicious emails correctly. A Confusion Matrix was further conducted to confuse the naïve Bayes test data.

Fig. 14 Classifier result

```
alpha            15.510010
Train Accuracy    0.979641
Test Accuracy     0.969549
Test Recall       0.777778
Test Precision    1.000000
```

Fig. 15 Confusion matrix

	Predicted 0	**Predicted 1**
Actual 0	1587	0
Actual 1	56	196

The above confusion matrix (Fig. 15) presented that a misclassification of 56 malicious emails as not- malicious emails whereas no misclassified for not-malicious emails.

Next the SVM algorithm was tested. The same procedure was repeated as the Naïve Bayes. The exact same reasoning was applied to Support Vector Machine model with Gaussian Kernel. Different training models were training, changing the regularization parameter C. A evaluation of the accuracy, recall and precision of the model with the test set (with the first 10 Support Vector Machine models & their metrics) (Table 4).

Furthermore, models which have Test Precision $\approx$ 1 were checked because the model which has the Test Precision $\approx$ 1 and has greater test accuracy then that model will be considered as the best model (Fig. 16).

Between these models with the highest possible precision, a selection was made to gain more test accuracy. The following model will be considered as the best Support Vector Machine model due to accuracy (Fig. 17).

The accuracy of the best Support Vector Machine Classifier was about 98% & it classified 85% of malicious emails correctly. This was a better result than Naïve Bayes. To test this further, Confusion Matrix was made for SVM Model (Fig. 18).

The above confusion matrix misclassified 37 malicious emails and only 1 not malicious email was classified. This was an impressive result.

Table 4 Test precision SVM

	C	Train Accuracy	Test Accuracy	Test Recall	Test Precision
0	500.0	1.0	0.979337	0.853175	0.99537
1	600.0	1.0	0.979337	0.853175	0.99537
2	700.0	1.0	0.979337	0.853175	0.99537
3	800.0	1.0	0.979337	0.853175	0.99537
4	900.0	1.0	0.979337	0.853175	0.99537
5	1000.0	1.0	0.979337	0.853175	0.99537
6	1100.0	1.0	0.979337	0.853175	0.99537
7	1200.0	1.0	0.979337	0.853175	0.99537
8	1300.0	1.0	0.979337	0.853175	0.99537
9	1400.0	1.0	0.979337	0.853175	0.99537

	C	Train Accuracy	Test Accuracy	Test Recall	Test Precision
0	500.0	1.0	0.979337	0.853175	0.99537
1	600.0	1.0	0.979337	0.853175	0.99537
2	700.0	1.0	0.979337	0.853175	0.99537
3	800.0	1.0	0.979337	0.853175	0.99537
4	900.0	1.0	0.979337	0.853175	0.99537

Fig. 16 Result for SVM

Fig. 17 Accuracy for SVM

```
C                    500.000000
Train Accuracy         1.000000
Test Accuracy          0.979337
Test Recall            0.853175
Test Precision         0.995370
```

Fig. 18 Confusion matrix SVM

	Predicted 0	Predicted 1
Actual 0	1586	1
Actual 1	37	215

	Multinomial Naïve Bayes	Support vector machine
Accuracy (%)	96	98
Malicious emails classification accuracy	77% (56 malicious emails misclassified)	85% (37 malicious emails misclassified)

Support vector machines produced better results due to the accuracy and the rate of misclassification in malicious emails. The Machine Learning algorithm successfully identified phishing emails and from the development and testing conducted, the support vector machine is the best classification algorithm to predict malicious emails accurately.

3.2   Critical Discussions

One of the objectives was to analyse the growth and extreme acceleration of phishing during the pandemic (COVID-19). This objective was met by highlighting and critically examining the facts with reasons to why this has happened. In-depth research has conducted on this topic to show how the pandemic of the virus (disease) was being used by attackers to their advantage. The cyber criminals made full use of the situation.

Working from home during the pandemic also accelerated the number of phishing attacks. People working from home become more vulnerable to phishing attacks as they were scared already. Social engineering was cleverly used by criminals and a lot of personal information was gained which also resulted in losses. Human errors are key to phishing as criminals want people to make mistakes so that it can be used against them.

Furthermore, different types of phishing attacks were analysed and studied such as spear phishing, smishing, whaling, etc. These are the specific and detailed advancements in phishing to target users in different ways. Spear phishing is one of the most successful methods. These attacks are still being developed even further with accuracy so that attackers are even more successful in their attacks.

The main machine learning algorithms were evaluated to conclude which is the best classifier such as logistic regression, decision tress, support vector machine and naïve bayes. This helped the author to examine which algorithm would be best fitted and tested for the purpose of detecting phishing emails.

Prevention frameworks which are currently in place were discussed to identify the gap in knowledge to that it can be fulfilled and researched. Preventing phishing structure has been provided by the author this in this report to reduce the phishing on users. Main aspects which were highlighted are that people need to be trained and need to be provided with as much awareness as possible because essentially phishing is mostly to do with human error. Common prevention tips have been given as well as giving the anti-phishing software's to prevent phishing.

Agile methodology helped the author to keep working in iterations so that the mistakes and errors can be corrected at early stages of the development.

Finally, the algorithm was developed in python using different machine learning techniques but the support vector machine model gave the most accurate results as it detected the malicious emails.

The theorem for Naive Bayes is based on the adjective Naïve which states that the data set is independent of each other. The occurrence of one attribute does not affect the chance of the other. Naïve Bayes will surpass the most efficient alternatives for small sample sizes. It is used in a variety of fields and is relatively stable, simple to execute, fast and precise.

For instance, recalling the kind of data that is required to solve and how the classification model wants to choose is decided by spam filtering in the email. Strong breaches and problems with non-classification of freedom may result in poor results. In reality, various classification models should be used for the same dataset, and therefore performance and device reliability considerations should also be taken into account.

The aim of a vector machine support algorithm is to find an N-dimensional space hyperplane (N the number of characteristics) that clearly categories data points. There are several potential hyperplanes to pick to differentiate the two types of data points. The aim at selecting an aircraft with the greatest range, i.e., the maximum distance from both data points. Maximizing the margin gap will improve potential data points so that they can be more securely identified. The has been successfully achieved.

The algorithm classifies malicious data successfully so therefore the algorithm has also been produced to the highest quality.

The research gave in-depth analysis of phishing in the pandemic and what the exact factors are which are contributing to phishing growth. Many recent articles and newspapers were read in regards to the issue. This helped the author understand where the problem lies and how it can be solved. Furthermore, social engineering is being utilised more than ever due to people working from home. The lockdown has stressed many people and individuals became more scared so whatever was posted regarding COVID-19 they would click and essentially, they would be the victims of Phishing.

This project has helped the author to raise awareness for phishing. The author has been creating articles which are being posted online via social media and other websites to raise awareness. People need to be careful where they click as it can be phishing related or can contain malware. The awareness the author is spreading has already gotten good feedback and people are starting to releasing and read more into this topic.

One of the techniques which was ignored during the project was a software development methodology called waterfall. This was ignored while created the Machine learning algorithm as it didn't seem to fit the project that well. The reason for this is because a stage needs to be 100% complete before moving on the next stage. This didn't fit the style and development of this algorithm for this project as many times the author had to go back to phases and improve on things. Furthermore, agile methodology was selected so that the author can work in iterations and developments can be conducted by coming back the previous phases.

An alternative hypothesis which could have been tested in this research was testing it on a real organisation. Once the algorithm would've been developed, an alternative test would've been testing it on larger data in an organisation to see how the algorithm does. This will also check the user's reactions and will give the opportunity for the author to make further improvements from a live environment.

The experimental results overall are complete and very reliable as they have been tested to the max with the training data set which was used. 98% precision is pretty good to detect phishing emails. Furthermore, in future this can be improved by testing it further with larger data and by using other algorithms with SVM.

4   Conclusions

In recent years the issue of phishing emails has become widespread. Phishing is an attempt to email victims to supply the users with sensitive information and to send them directly to the photographer. It is necessary to identify them. There are many methods for detecting Phishing emails, but there are some drawbacks, such as poor accuracy, information that cannot be detected, and a low detection rate, which can be similar to legitimate emails.

In this study, the precision of the detection of phishing email was evaluated on the basis of manual selection of function and automatic selection of function on five algorithms. Given a set of words, feature selection was used to obtain words that allow us to distinguish between malicious and not-malicious emails.

Various classifiers were compared to check the accuracy in predicting the class attribute. The Support Vector Machine (SVM) method gives the highest classification accuracy & the rate of misclassification in malicious emails than the Naïve Bayes method. 98% accuracy was achieved but this can still be further increased if tested on large corpus of training data.

In future work, the study is not enough and there is room for improving the detection of malicious emails. As functional filtering methods must be further developed to counter the continuous growth over time of the phishers' new techniques. Therefore, the author proposes a new automated method to retrieve new functionality from new raw emails to increase phishing email identification accuracy and to manage the growth of the phisher technology.

Deep Neural Network (DNN) can be used to modify the proposed solution that was used as the detection model. DNN is a system for the learning of machines and can model complex non-linear interactions by the use of different layers. First of all, the highest performing functionality with problems that outperform other methods is the core advantage of using DNN. Secondly, it decreases the need for the engineering of phishing emails which is one of the most critical activities. Finally, it is reasonably straightforward to apply DNN's architecture to new issues. However, the DNN has some drawbacks which result in the thesis not being used. DNN is highly costly to prepare computationally.

Once the phishing email identification mechanism has been developed for phishing, the proposed architecture may be used to improve the current phishing system. A dataset used to construct the malicious email detection model plays a significant part, so any newly usable dataset can be used to automatically boost the detection model by applying it to the offline date range.

From binary classification (phishing and ham emails) to a ternary classification model which can be categorised into spam, phishing and ham emails may be expanded from the same detection model. This upgrade can be achieved by introducing further functionality representing spam and upgrading the dataset to include spam emails instances. Eventually, you can add additional functionality to the application list to improve the pre-processing algorithm which will enhance the system's ability to explore new actions using the advanced FEaR algorithm. Finally, certain changes may be considered in future research, whether to look more closely at a single phishing email property or expand the classification model to wider domains.

References

1. Abu-Nimeh S, Nappa D, Wang X, Nair S (2007) A comparison of machine learning techniques for phishing detection. In: Proceedings of the anti-phishing working groups 2nd annual eCrime researchers summit, pp 60–69
2. Abu-Nimeh S, Nappa D, Wang X, Nair S (2009) Distributed phishing detection by applying variable selection using Bayesian additive regression trees. In: 2009 IEEE international conference on communications. IEEE, pp 1–5
3. Adida B, Bond M, Clulow J, Lin A, Murdoch S, Anderson R, Rivest R (2006) Phish and chips. In: International workshop on security protocols. Springer, Berlin, Heidelberg, pp 40–48
4. Alguliev RM, Aliguliyev RM, Nazirova SA (2011) Classification of textual e-mail spam using data mining techniques. Appl Comput Intell Soft Comput 2011:10
5. Almomani A, Gupta BB, Atawneh S, Meulenberg A, Almomani E (2013) A survey of phishing email filtering techniques. IEEE Commun Surv Tutor 15(4):2070–2090
6. AmtrustFinancial (2021) Social engineering scams rise during COVID-19 | AmTrust Financial. https://amtrustfinancial.com/blog/small-business/social-engineering-scams-rise-covid19-pandemic. Accessed 12 Jan 2021
7. Azad MA, Morla R (2011) Multistage spit detection in transit voip. In: SoftCOM 2011, 19th international conference on software, telecommunications and computer networks. IEEE, pp 1–9
8. Basnet RB, Sung AH (2010) Classifying phishing emails using confidence-weighted linear classifiers. In: International conference on information security and artificial intelligence (ISAI), pp 108–112
9. Bergholz A, Chang J. Paass G, Reichartz F, Strobel S (2008) Improved phishing detection using model-based features. In: CEAS
10. Brewster T (2021) Coronavirus scam alert: watch out for these risky COVID-19 Websites and Emails. [online] Forbes. https://www.forbes.com/sites/thomasbrewster/2020/03/12/corona virus-scam-alert-watch-out-for-these-risky-covid-19-websites-and-emails/#2f558bca1099. Accessed 4 Jan 2021
11. Cao Y, Han W, Le Y (2008) Anti-phishing based on automated individual white-list. In: Proceedings of the 4th ACM workshop on digital identity management, pp 51–60
12. Gansterer WN, Pölz D (2009) E-mail classification for phishing defense. In: European conference on information retrieval. Springer, Berlin, Heidelberg, pp 449–460
13. Jameel NGM, George LE (2013) Detection of phishing emails using feed forward neural network. Int J Comput Appl 77(7)
14. Khonji M, Jones A, Iraqi Y (2013) An empirical evaluation for feature selection methods in phishing email classification. Int J Comput Syst Sci Eng 28(1):37–51
15. Kumar RK, Poonkuzhali G, Sudhakar P (2012) Comparative study on email spam classifier using data mining techniques. In: Proceedings of the international multiconference of engineers and computer scientists, vol 1, pp 14–16
16. Kumaraguru P, Sheng S, Acquisti A, Cranor LF, Hong J (2010) Teaching Johnny not to fall for phish. ACM Trans Internet Technol (TOIT) 10(2):1–31
17. Ma L, Torney R, Watters P, Brown S (2009) Automatically generating classifier for phishing email prediction. In: 2009 10th international symposium on pervasive systems, algorithms, and networks. IEEE, pp 779–783
18. Muncaster P (2020) COVID19 fears drive phishing emails up 667% in under a month. [online] Infosecurity Magazine. Available at: [Accessed 9 June 2020]

19. Nizamani S, Memon N, Glasdam M, Nguyen DD (2014) Detection of fraudulent emails by employing advanced feature abundance. Egypt Inform J 15(3):169–174
20. Paaß G, Bergholz A (2009) AntiPhish-machine learning for phishing detection. Project Exhibition at ECML/PKDD, 8
21. Ramanathan V, Wechsler H (2012) phishGILLNET—phishing detection methodology using probabilistic latent semantic analysis, AdaBoost, and co-training. EURASIP J Inf Secur 2012(1):1
22. Thomson Reuters Institute (2021) COVID-19 and financial scams, fraud and misinformation: what you need to know—Thomson Reuters Institute. https://www.thomsonreuters.com/en-us/posts/government/covid-19-scams-frauds/. Accessed 4 Jan 2021
23. Tidy J (2020) Google blocking 18M coronavirus scam emails a day. *BBC News.* [online] Available at: [Accessed 7 June 2020].
24. Toolan F, Carthy J (2009) Phishing detection using classifier ensembles. In: 2009 eCrime researchers summit. IEEE, pp 1–9
25. Wu Y, Zhao Z, Qiu Y, Bao F (2010) Blocking foxy phishing emails with historical information. In: 2010 IEEE international conference on communications. IEEE, pp 1–5
26. Zhang W, Lu H, Xu B, Yang H (2013) Web phishing detection based on page spatial layout similarity. Informatica 37(3)

Artificial Intelligence, Its Applications in Different Sectors and Challenges: Bangladesh Context

Kudrat-E-Khuda Babu

Abstract Bangladesh is home to 160 million people and is the most densely populated country in the world. Once, this country was dependent only on agriculture. But with the change of time, the advent of advanced technology has taken place in every field of this country. Now Bangladesh is being transformed into a modern technology-based country. Automation and control technology is being applied in various industries. Artificial Intelligence (AI), Internet of Things (IoT), Big Data, Blockchain, etc. have become very popular in Bangladesh. Although late, the impact of these technologies has begun to be felt in different sectors of the country. For the effective implementation of AI, several specific sectors such as services, transportation, education, agriculture, health, and the environment have been identified in Bangladesh. Overall in Bangladesh, we see extensive use of AI technologies, for example, ride-sharing, natural language processing (NLP) for Bengali, ChatBots, booking hotels, buying air tickets and real-time mapping, etc. As about 34% of youth are technology-driven in the country now, the successful integration of AI technologies will lead Bangladesh towards a prosperous future. For implementing AI technology, the government of Bangladesh should have to undertake huge preparation. So there will be many challenges if the technology is adopted without any proper preparation. This challenge is not only for Bangladesh to adopting AI, every country has to go through these issues. So the Bangladesh government needs to have preparation such as infrastructural development and technology enhancement and some other relevant issues to control the AI. This chapter briefly discusses the current technological situations in Bangladesh along with the concept of AI, its applications in different sectors, and the relevant challenges. The chapter also suggests the steps to be taken in the use of AI technology to maintain certain aspects such as policy, information privacy, security, and regulations.

Keywords Artificial intelligence (AI) · Bangladesh · Economic growth · Fourth industrial revolution (4IR) · Information technology

K.-E.-K. Babu (✉)
Daffodil International University, Dhaka 1207, Bangladesh

R. Montasari and H. Jahankhani (eds.), *Artificial Intelligence in Cyber Security: Impact and Implications*, Advanced Sciences and Technologies for Security Applications, https://doi.org/10.1007/978-3-030-88040-8_4

1 Introduction

Artificial Intelligence (AI) refers to the ability of a machine to perform tasks based on human reasoning -such as reasoning, observation, learning, critical thinking, and basic leadership. Initially, it was invented to mimic human insights, but now AI has been developed in a way that goes far beyond its previous capabilities. As a result of tremendous advances in data collection, management, and computing control, it will now be able to control various assignments, strengthen the network, and improve profitability. As the scope of AI's capabilities has expanded significantly, so has the number of different areas of its usefulness. Bangladesh's government has adopted the AI system for the digitization of the country nationally. The process of digitization of the country has started a decade ago. Now AI is acting as the accelerator of the process.

Bangladesh's Prime Minister Sheikh Hasina has already announced, "Five G (5G) will be on run within 2023. Future technologies like artificial intelligence, robotics, big data, blockchain, and IoT will be widespread." Bangladesh is committed to moving forward along that path. Information Communication and Technology (ICT) has been adopted as a tool for sustainable development to transform Bangladesh into a technologically advanced country in the next decade. Bangladesh's export revenue from ICT was only $26 million, which has now reached about $1 billion. Bangladesh has already invested largely in the ICT sector. The country has undertaken different big projects including 16 Hi-Tech Park, 7 (Seven) Technology Park, 12 IT Training and Incubation Center, and Tier-IV Data Center [7]. The touch of technological development of Bangladesh is also felt in space. In 2016, the country launched its first satellite Bangabandhu-1 into space. With $1,47,570$ km^2 area, Bangladesh has a population of over 163 million, of which about 40 million are students. As a result, the country is facing significant challenges over the issue of employment. Many people will lose their job opportunities for AI in the country. AI will replace human skills. As a result, it will be difficult for the countries like Bangladesh with a large number of low-skilled people to survive. Meanwhile, according to the World Economic Forum, AI and robots will create far more workplaces than displacement. Due to AI and its related technologies, economic growth will increase faster than ever before and create more job opportunities [10]. Besides, the fourth industrial revolution is likely to create 133 million new roles. As the country has integrated the 2030 Agenda in its seventh Fiscal Year Plan (2016–2020), so this is a great chance to execute the agenda in light of the needs of SDGs. The government, NGOs, philanthropists, tech companies, and organizations of the country, those who collect or generate a significant amount of data, will have to play a decisive role to expedite the process. Some issues need to be addressed first regarding this. There is a big challenge for accessibility of data and a shortage of experts or resource persons who can improve the capacities of AI, develop models, and address the possible arising issues. AI will have significant roles in addressing the challenges of SDGs. McKinsey Global Institute has identified about 160 challenges to the SDGs where AI can play a vital role to address the issues. Bangladesh has adopted a policy to address the challenges of SDGs using artificial intelligence.

2 Artificial Intelligence: General Meaning and Concept

Artificial Intelligence (AI) is a technology that is capable of performing human-like intellectual tasks. For example, AI can perform intellectual functions, such as visual perception, speech, recognition or decision making, and language translation. It has revolutionized the world in information technology. It is an advanced program of computer science that can work and respond like a human. AI has replaced almost all the functions of human life and changed the type, nature, and speed of work. In a word, it can work as a competitor and alternative to human beings. Even, in some fields, such as employment, economics, communications, war, privacy, security, ethics, healthcare, etc., it has shown greater efficiency than humans. However, we are not yet sure about its vast impacts and we have not yet seen its future evolution whether it will help make the world more conducive to human survival or become a future threat to humanity. AI is the production of a computer model of human intelligence processes which include—learning (the acquisition of information and rules for using the information), reasoning (using rules to reach approximate or definite conclusions), and self-correction [12]. An AI is comprised of these particular applications—expert systems, speech recognition, and machine vision. However, day by day, the current state of artificial intelligence is being changed swiftly and its evolution is accelerating. The research for developing an AI has started decades ago. American scientist John McCarthy first devised AI in 1955. So, he is also considered the co-founder of this sector. The latest form of AI is an outcome of the continuous devotion and hard works of scientists and researchers. Now, we can store data torrents remotely through cloud computing. Moreover, it is not so expensive to know about neural network technology nowadays. Even the largest tech industries, such as Google, Facebook, Microsoft, and IBM are focusing on AI research and they see a huge potential there. The goal of AI is to create expert systems- which demonstrate intelligent behavior, learn, demonstrate, explain and advise intelligent behavior to its users, and apply human intelligence to machine building that understands, thinks, learns, and behaves. Generally, it can be said that AI is a combination of science and technology based on mathematics, psychology, computer science, biology, engineering, and linguistics. Usually, there are three categories of AI. These are:

(a) Narrow: Narrow AI (sometimes called "weak AI.") is focusing on executing a single task but it has limitations of interaction. Checking weather reports, controlling smart home devices, or giving us answers to general questions pulled from the central database are some of the examples of narrow AI.

(b)　General: We are still in Narrow AI-but scientists believe they are making progress towards general AI. It learns from experience and can understand the data and make a decision based on data.

(c)　Super: In the near future, AI may become intellectually superior to humans in every way. AI robots would probably have a problem-solving attitude, accomplish awareness and work with no human association, maybe at the directions of another AI [8].

3　Digital Bangladesh and Artificial Intelligence

Bangladesh is working tirelessly to keep a successful footprint in the world as one of the best performers in Information Technology (IT) and Information Technology Enabled Services (ITeS). This effort will make the South Asian court one of the most digitally developed countries in the world by the coming years, which is also one of the basic commitments of the current government [13]. With this vision of "Digital Bangladesh", many renowned organizations like JP Morgan, Goldman Sachs, and Gartner have recommended Bangladesh as a great example of the future of IT and ITeS. The theme of Digital Bangladesh reflects the idea of integrating modern technologies in every sector of the country. Its major aim is to use technology in communication, education, health, training, transportation, administrative and bureaucratic services, social services, literacy, electricity, wireless, internet coverage, social media services, e-services, access points, policymaking, agriculture, industry, and commerce and accelerate the development of the country. Digital Bangladesh refers to a digitally developed society that ensures an ICT-driven knowledge-based society where everyone has easy access to online information and services. These services will be instantly accessible on the web and mobile, where the administration, semi-public and private sectors will be ready to take advantage of the latest innovations. Bangladesh's government has an intention to take the advantage of AI in its journey towards "Digital Bangladesh", a digitally advanced country.

4　How Bangladesh Can Gain Economic Benefits from AI?

With the extremely dynamic economy, Bangladesh has become the tiger of Asia. Since the last decade, this country has been showing its economic strength with vibrant and continuous GDP growth and the digital revolution and renovation. Now Bangladesh has been considered as the pioneer among the developing countries for e-government services like e-Govt. and e-Citizen. According to GNI, this country's economy has been ranked as the 42nd largest economy in the world and as per the ranking by Public-Private Partnership (PPP) it stood at the 31st position. Bangladesh has been included in the next eleven emerging markets and frontier five. Bangladesh's economy is the second-fastest-growing economy in the world with a growth rate of

7.1%, and the impact of AI's could increase the country's expected economy by 45%, according to the IMF. Some of the most common taxonomies of AI invention are natural language processing, virtual assistants, computer vision, and robotic process automation. The role of AI in business is a fascinating one for the global economy. AI is anything but a solitary innovation yet a group of advances. Some general classifications of AI innovations: Natural Language Processing, virtual assistants, Computer Vision, and Robotic Process Automation [3]. The role of AI in business and the worldwide economy is an interesting issue. Bangladesh is one of the fastest-growing economies in the world as a result of significant GDP growth (8.13 FY19). Now the question is, how did Bangladesh adopt AI and become a fast driver of IIR and was able to bring a rapid change in the socio-economic condition of the country? The right strategies and actions for digital transformation through AI in citizen services, manufacturing, agriculture, health, mobility and transportation, finance and trade, and all other possible areas can dramatically boost Bangladesh's economic growth. Specialized skills and a vibrant start-up ecosystem need to be developed to further expand Bangladesh's economy and keep pace with the stimulus and pace of socio-economic development. AI will work through work skills and development, driving development through intelligent automation, human–machine collaboration, and promotion of innovation [15]. Part of that is now playing a role as a smart factory where people and machines are working one after another to improve results. In the insurance sector, machines work monotonously to enable people to focus on continuous mind-boggling, judgment-based preparation and client administration. New opportunities are similarly created through the development in the process of AI. For example, thanks to AI, Google Maps contributes to the growth of ridesharing platforms like Uber, Shohaz, and Pathao, transforms individual transport, and creates jobs for a large number of people in Bangladesh. As artificial intelligence is used in robotics, big data research, the IoT, and genomics, the breadth of its usefulness has also reflected on the lives of farmers and improving their living standards [4, 5]. For example, AI has been used to ascertain the unpredictability of climate change or soil conditions, or it has reduced the widespread cost of agricultural labor in many countries across the world. AI has also been contributing to the development of the agriculture sector worldwide by increasing crop yields, reducing farmers' costs, and increase profits. However, the widespread use of AI has been started for the last several years through the concept of advanced technology was developed a decade ago. In the coming days, AI technologies will bring changes to the process of manufacturing the products and their marketing process. If we want a sustainable business ecosystem, then the use of advanced technology in the whole process is unavoidable. AI technology will ensure that people get adequate support in education and opportunities for their livelihood. Global Research Institutes claim that Bangladesh's economy will double in the next ten years and that Bangladesh will emerge as a model for developing countries at some point through continuous innovation.

5 AI and Job-Field Economic Growth for Bangladesh

Creating employment opportunities for a population of 163 million and more than 40 million students in a country of just 147,570 km^2 area is a huge challenge. Due to the increasing intervention of technology, the employment opportunities of the people are constantly shrinking. In that case, there is a risk of increasing the possibility of unemployment. So it is very important to determine effective strategies to deal with this loss threat. If 4IR technology will be adopted in the industry, many opportunities for employment will be closed [16]. So some people will lose their jobs and those whose jobs will survive may need to be reconsidered. If we can develop capacities in technologies and equipment, then the production cost will be lower compared to the international market. And with the advancement, the local companies will get the opportunity of making more capital and help them expand their periphery and create the opportunities to absorb the employees who will be at risk of losing jobs. But, the manufacturing strategy of Bangladesh is so far as to import capital machinery and technologies with the involvement of local laborers. As a result, the import duty rate on capital machinery has been set lowest one percent which is an impediment to create knowledge-intensive jobs in Bangladesh. Despite this challenge, Bangladesh needs to compete in the global market and develop its capacity through technological innovation in manufacturing. Meanwhile, the World Economic Forum said AI and Robots would create more jobs than they displace and the economic growth will boost up than ever before due to AI and related technologies.

Besides, The Fourth Industrial Revolution is likely to create 133 million more new jobs though there is an apprehension that around 75 million jobs may be displaced by 2022. There is an analysis on how AI can both destroy and create jobs through the displacement and income effects (this is a simplified analysis—in practice, there will be a more complex range of economic effects at work). PWC Statistics shows, most of the emerging technology from steam engines to computers, displaced some existing jobs but also created new jobs and large productivity gains [11]. According to PwC, to "displacement the effect" of AI, the government should invest more in STEAM—(Science, Technology, Engineering, Art and Design, and Mathematics) education. In this perspective, the government should encourage workers to update and adapt their skills with the new technology continuously. Higher academic graduates should be engaged in next-generation capital machinery and change the country's labor-intensive manufacturing strategy to knowledge-intensive ecosystems.

6 Fourth Industrial Revolution (4IR)

The Fourth Industrial Revolution (4IR) through its advancement in the Internet of Things (IoT), cyber-physical systems, and the Internet of systems has changed our lives, works. Adopting the new technologies in workplaces, inter-connected machines are interacting and visualize the entire production chain. It also helps us make decisions faster and independently. This industrial revolution has already impacted our society, economy, and industry. It is considered as the extension of the 3rd Industrial Revolution, commonly known as the Digital Revolution [1]. However, the fourth industrial revolution is going on in every country. Influencing every industry in the world and bringing huge changes at an extraordinary pace. The first industrial revolution took place in the eighteenth and nineteenth centuries as a result of steam engines and other technological advances. At that time there was a great change from the agrarian society to wider industrialization. Later, the Second Industrial Revolution came with the invention of electricity. This marked the beginning of a wide range of industries and large-scale production as well as technological advances. The Fourth Industrial Revolution began with the discovery of artificial intelligence. Everything in the world is constantly changing. The technologies of this industrial revolution, such as genome editing, 3-D printing, and robotics with artificial intelligence, are changing almost everything related to human life, including art, communication, medicine, research. An important point here is that this revolution will be conducted in terms of human choice. New technologies are evolving at an increasingly rapid pace and are being applied that make an impact on people's identities, communities, social and political structures. As a result, how we engage with the technologies of the Fourth Industrial Revolution is manifested by our commitment to each other, our opportunity for self-realization, and our ability to positively influence the world. At the same time, we must not only enjoy the benefits of this revolution, but we must also take appropriate responsibility for the purpose and structure of this revolution.

7 AI for Different Sectors of Bangladesh

Bangladesh is the most densely populated country in the world. Once, this country was dependent only on agriculture. But with the change of time, the advent of advanced technology has taken place in every field of this country. Currently, Bangladesh is being transformed into a modern technology-based country. Automation and control technology is being applied in various industries. AI, IoT, Big Data, Blockchain, etc. have become very popular in Bangladesh [2]. Although late, the impact of these technologies has begun to be felt in different sectors of the country. For the effective implementation of AI, several specific sectors such as services, transportation, education, agriculture, health, and the environment have been identified in Bangladesh. Overall in Bangladesh, we see extensive use of AI technologies, for example, ride-sharing, natural language processing for Bengali, ChatBots, booking hotels, buying air tickets and real-time mapping, etc. As about 34% of youth are technology-driven in the country now, the successful integration of AI technologies will lead Bangladesh towards a prosperous future.

7.1 AI Envisioned National Priorities of Bangladesh

Through research and development strategy planning, AI technology can be safely used for the betterment of all levels of society and for the development of the world to come. Bangladesh has received the final recommendation to graduate from the least developed country (LDC) category at the second triennial review by the Committee for Development Policy of the United Nations (UN-CDP) held during February 22–26, 2021. With this, Bangladesh has once again met all three criteria—per capita gross national income (GNI), human assets index (HAI), and economic vulnerability (EVI) index—to graduate from the LDC group. If things go well, Bangladesh is expected to graduate from the LDC in 2026. In this context, the use of artificial intelligence in the country's economic, research, industrial, agricultural, and medical fields will help Bangladesh become the fastest-growing country in South Asia [9]. In this case, we have set 8 national priority sectors – AI for public service, manufacturing, agriculture, smart mobility and transportation, skill and education, finance and trade, and health. These priorities have been set as per the GDP contribution and basic needs for people. More priorities would be set for the inclusion of more industries under the concept of artificial intellectual AI + X (Anything) in phases. Thanks to the further advancement of AI, artificial intelligent technology can lead to financial prosperity as well as prosperity in almost all spheres of society including health, agriculture, and education.

7.2 AI for Public Service Delivery

Utilizing the AI, Bangladesh has already developed *Eksheba* for the citizens of the country which is serving as a one-stop service access point for all services of the government. Now the citizens can get access to any online services by using a single identity. The government of Bangladesh has undertaken an initiative to analyze all the services provided by its various offices and has identified more than 2700 online services. According to the Bangladesh Economic Profile 2018, the service industry contributes 56.5% of the total GDP to the economy. Almost all government offices can use artificial intelligence applications. The most commonly used of the existing services of the government, respective individuals usually search for various information and the use of topics related to their respective information. The use of AI technology can reduce the administrative distance between government and citizens in getting to access all government services. This will increase the quality of work as well as increase the pace of overall development. So with an aim to widespread use of artificial intelligence, different countries are scrutinizing AI-based applications around the world.

7.3 AI for Manufacturing

The growth for Bangladesh's manufacturing sector has been projected to grow by 13%. According to the observations of many concerned people, this growth rate will increase further. Over the past four years, the large and medium production index has been growing at an annual rate of 11%. As a result, in the first four months of the financial year, the output of large and medium industries was increased by 20% over the previous year. As per the estimation of Bangladesh Economic Profile 2018, only the manufacturing industry contributes 29.2% of the total GDP in the country's economy. AI and machine learning have led to the fourth industrial revolution which has taken a toll on every industrial organization in the world. Data analysis and predictive decisions through this new technology in the manufacturing sector will reduce the raw materials in the manufacturing sector of the industry, increase efficiency and make the supply chains more favorable [14]. Smart manufacturing usually includes overall equipment functionality (OEE), custom, and adaptive manufacturing. Despite adopting the new technologies, some impediments interrupt the process of maximizing growth in the manufacturing sector. But there is nothing to be worried about at all. To solve the complications, AI will be enabled for helping to decide by analyzing the predictive demand and supply system alongside developing the appropriate skills of the workers.

7.4 AI for Agriculture

Once, agriculture was the mainstay of Bangladesh's economy. Eighty percent of the people in the country were involved in this industry. Agriculture still plays an important role in the development and strength of Bangladesh's economy. Currently, more than three-quarters of the total population in rural areas make their living from agriculture. Bangladesh became a food self-sufficient country in 2009. So far, the government has undertaken huge activities for the welfare and development of the agriculture sector. It has already established 245 agriculture information centers to help farmers to grow crops across the country. Bangladesh Bank, the central bank of the country, in an initiative to open bank accounts for farmers by deposing only Tk. 10 only ($0.124). Besides, the government has introduced an online service – krishi.gov.bd portal—and a hotline (3331) to support the farmers. The agricultural industry contributes 14.2 percent of the total economy, as per the information of Bangladesh Economic Profile 2018. In his sector, many technology-based activities are usually done. For example, to know the actual condition of crop fields, activities analysis of cows, feeding behavior of fish are being done by the AI and IoT at the digital agriculture platforms. AI also mapping the crop stage using the satellite image, hydroponics & vertical agriculture. It also helping factories to move towards building plant factories and producing big data for prediction mapping. Thus, AI technology accelerates the development of the industry by regularly coordinating with various plans and development programs. Proper application of artificial intelligence technology can address a number of challenges such as dynamic soil topology maps, disease forecasting team systems for single crops, crop prediction automation, image-based disease recognition, and health monitoring.

7.5 AI for Smart Mobility and Transportation

There has been a huge development in the communication sector of Bangladesh. Anyone can seek the country's communication revolution through the reflection of its roads, bridges, and flyovers. This South Asian country has a total of 3,813 km highways, and 368.62 km four-lane highways, and there are 12, 91,707 m bridges/culverts till 2016 [6]. The achieve Bangladesh's vision 2021, the current government, which is in the power for around 12 years, has taken a number of mega projects such as Padma Multipurpose Bridge, which is expected to complete by 2022 and transform the lives of 30 million southern people, Payra Deep Seaport (country's third seaport), Dhaka Metro Rail, Dhaka-Chittagong elevated expressway, Dhaka elevated expressway, Karnafuli underwater tunnel, Bus rapid transit in Dhaka, etc. The construction of most of the projects is underway. So, AI can a bigger role in analyzing data on the road, vehicles, driver behavior, vehicle behavior, speed limit, turns, speed breakers, weather, and infrastructure. This technology can also monitor untoward incidents on road like robbery, can record transport in apps to avoid harassment, activities of law enforcers.

7.6 AI for Skill and Education

Bangladesh has made a wonderful success in the education sector over the past decade especially ensuring access of girls to every level of education. The enrollment rate at the primary school level education has increased from 80% in 2000 to 100% in 2015, according to a report of the World Bank. The adult literacy rate of the country, which has already achieved gender parity in primary and secondary level education, is 72.3% up to the year 2016. The government has been distributing over 2.3 billion free books on the first day of every year since 2011. Alongside undertaking many projects for encouraging the students to educate and decreasing the dropout, the government has been giving scholarships and stipends to at least 17 million students out of a total of around 37 million students in Bangladesh. The government has set multimedia classrooms at 32,000 educational institutes across the country aiming to transform the teaching system [4]. The country is considering education as one of the core strategies to alleviate poverty and facilitate development. The government has already taken huge steps to encourage the students to ICT, raise their skills by introducing ICT courses from the school level education. It has also opened many ICT-based universities, labs in the country and it has been establishing ICT parks to facilitate the ICT industry. The introduction of AI at every level of education and the establishment of ICT parks and labs can boost up the country's ICT sector more than the imagination. AI can contribute to developing skills, capacity building of teachers, and many other issues related to the development of the country's growth.

7.7 AI for Finance and Trade

Artificial Intelligence (AI) technology has been adopted in the country's financial services industry in the early 80 s. It is one of the sectors of the country where technology has been introduced first. This is a sector that produces significantly larger data sets due to the complexity of the markets than other sectors. So according to customer experience and skill requirements, financial services in this sector are more interested in adopting emerging technologies than others. People usually make decisions in their business based on their inherent and extensive knowledge of the economy. But the range of this sector is so wide that it is not possible to solve it quickly with the common sense of the man in various complex calculations in different financial institutions and in analyzing a large amount of data. Therefore, the incorporation of information technology (AI) to manage financial transactions in the banking system has revolutionized the sector. According to the Bangladesh Bank (BB) guidelines, "Commercial banks are categorized as Category-1 and Category-2 Category-1 means Centralized ICT Operation for managing core business application solution through Data Center (DC) with backup assets for the continuation of critical services including Disaster Recovery Site (DRS)/Secondary Data Center to which all other offices, branches, and booths are connected through WAN with 24/7 attended operation. Category-2 stands for decentralizing ICT operation for managing distributed business application solutions hosted at DC or operational offices/branches with backup assets for the continuation of critical services connected through WAN or having the standalone operation. NLP Bot-based RSD will save work hours/costs/eliminate tech/education divide, AI-based credit management will eliminate fraud/enhance credit availability/enhance economy, centralized KYC automation-improve service dealing, RPA in trade and e-government-will eliminate duplicate, AI for G2B single point service delivery".

7.8 AI for Health

Bangladesh's healthcare sector has always failed to provide quality healthcare. Due to lack of quality medical equipment, lack of research in medical science, old practice in medical services, incompetence of health workers and lack of necessary knowledge of doctors and lack of use of advanced technology, the healthcare service in Bangladesh is very low compared to other countries in the world. To overcome this problem, the role of adequate research as well as advanced technology is very important. Due to the lack of quality medical services, the people of Bangladesh are suffering from the most common non-communicable diseases like diabetes, heart attack, hypertension, and liver disease. Since over 60% of the expenses of healthcare are usually borne by the patients, the poor and helpless people are deprived of this conventional healthcare due to a lack of money. There is a projection that Bangladesh has to be the seventh-largest diabetic population across the world by 2030. According to the World Health Organization (WHO), there are an esteemed 3.05 physicians per 10,000 population and 1.07 nurses for 10,000 population (estimates based on MoHFW HRD 2011). So, there is no substitute for the ongoing consequential use of AI in the rapidly improving services of healthcare information and the vast access to huge data scientific strategies. Health AI networks can solve many problems in the healthcare sector by providing quality services such as centrally connected health registries, wearable IoT devices, decision support systems, portable healthcare.

8 Legal Policy Framework for Using AI

The steps to be taken in the use of AI technology to maintain certain aspects such as policy, information privacy, security, and regulations are:

- First, we need to determine the process of using AI-such as when, where, and how to use it.
- A long-term analysis of AI development is needed to help determine long-term damage to society. For example, Stanford University conducted a study called 'One Hundred Years of Artificial Intelligence Study'.
- Several AIs need to be connected in such a way that they can create checks and balances on their own, test each other, and act as a reliance network for overall decision making.
- Operational ethics committee should be formed in the interest of the progress of robotic research. Besides, ethical charters must be developed for the further development of AI research.
- Further development of AI technology and the consequent impact on human employment and the economy, or new challenges that should be addressed by government agencies need to be accelerated.

9 The Challenges

For implementing artificial intelligence technology, the government of Bangladesh should have to undertake huge preparation. So there will be many challenges if the technology is adopted without any proper preparation. This challenge is not only for Bangladesh to adopting AI, every country has to go through these issues. So the Bangladesh government needs to have preparation such as infrastructural development and technology enhancement and some other following issues to control the AI.

9.1 Accompanying the Transformation

Artificial Intelligence (AI) is not just a technological invention, it is also social innovation. Therefore, the impact of artificial intelligence can come along with various benefits as well as various complexities that can completely transform society, including the public sector. The application of AI everywhere can often be a hindrance to the relationship between users and organizations. So society and cultural change must be brought about to create a smooth and sustainable relationship between digital technology, technology users, and administration to overcome barriers.

9.2 Data Eco-System

Information is called the raw material of artificial intelligence technology. And so there needs to be high quality and interconnected full data bank to supply the raw materials of artificial intelligence technology as well as all the tools and strategies of AI to create interactive and smart systems. In the case of data marketing in Bangladesh, the lack of access to data is identified as a major barrier. Not all administrative information of government ministries and departments is available or collected online. In that case, it is necessary to have available data available for research, business, and decision making. In this case, data.gov.bd acts as an open data platform for collecting, producing, and managing the necessary data.

9.3 Technology and Infrastructure

There is a lack of technological development as well as infrastructural development for adopting AI technology in Bangladesh. There is a great lack of infrastructure development in the areas of data handling, storage, computing, scaling, extensibility, and data security. In that case, the lack of infrastructural development remains a challenging issue for the implementation of this technology. Therefore, technologies like big data, machine learning, deep learning, and decision making are not widely available in Bangladesh. So to tackle these challenges we need to create a cloud platform for AI technologies. In addition, high counting equipment must be installed and a responsible pool of training must be provided to make them effective.

9.4   Skilled AI Resources

There is a dearth of skilled manpower in Bangladesh for the implementation of artificial intelligence technology. However, the Bangladesh government has already taken extensive measures to build skilled manpower. Bangladesh has already started many IT training projects under various projects for the use of this technology by public sector professionals. In this case, three or six months of training can be arranged initially for those who will be responsible for the application of technology by different governments. In this way, the manpower of the whole country can be brought for training in phases.

9.5   Connectivity

AI technology and the Internet of Things (IoT) require high-speed networks. In that case, at least 5G network is required which is still a challenge for Bangladesh. There is another challenge to establish wireless interconnection among LoRa, Sigfox, or forms of Narrowband IoT network due to low power IoT devices. Bangladesh is expected to launch a 5G network by 2023.

9.6   Economic Impacts: Inequality and Technological Unemployment

While many people around the world fear that the technology could cause massive job losses, the World Economic Forum says AI and robots could create more jobs than displacement. So if people are fired then the government has to create alternative fields for them. And if it creates new job opportunities, then skilled people have to be created accordingly. The number of lower-end workers is one of the major contributors to our economy in various manufacturing sectors. If they lose their jobs due to automation, the government must make alternative arrangements for their livelihood. Large-scale population training should be provided free of cost or with subsidy.

9.7   Accountability, Transparency, and Privacy

In the case of artificial intelligence technology, the people involved need to have the same skills as transparency in every aspect of the technology. Failure to do so could lead to a decline in the sector. So there is no clear way to know who will be responsible for any unwanted disaster. Because no one can say exactly why an algorithm relies on a DIP system in a particular decision-making process. The EU General Data Protection Regulation (GDPR) may be a solution to this situation but each process should have an explanation.

9.8   Human Dignity, Autonomy, and Psychological Impact

When machines lose control of various aspects of human daily life, it is feared that at some point, human roles will continue to lose importance. Until the end, it is not clear how this will affect human relationships and the ecosystem of work if they become progressively indifferent to natural languages.

9.9   AI Safety

Poor design of AI systems can disrupt the protection of artificial intelligence which can lead to any kind of accident or any harmful and inappropriate use of AI. AI frameworks need to maintain a strategic distance from accident problems.

9.9.1   Legal and Ethical Framework

Artificial Intelligence has constantly influenced every aspect of society, improving the process of various institutional work. It has already brought a revolution in banking and financial institutions, law enforcement agencies, and the healthcare system. In this case, the decision-making process should have a strong legal and ethical framework on how AI applications will work. AI's policies should be appropriate, fundamentally flexible, evaluable, and workable.

10   Conclusion

How AI technology is taking over human intellectual functions, it is very difficult to predict where the future of humanity will stand in the days to come. We can only accept AI when it will only have a positive impact on the welfare of society and humanity. However, any technology can have both positive and negative effects on society. So one must be prepared to face the negative effects while applying this technology. Moreover, we must have a strong legal framework in place to reduce the challenges associated with AI and to compensate the aggrieved parties in the event of a serious error. Moreover, the proper implementation of that law can reduce the serious threat to humanity. The success, AI technology has brought so far has led to much enthusiasm among policymakers around the world for further development of this technology. Due to differences in capabilities, they do not have the same level of strategic priority in this competition. Interested organizations are developing new strategies in their respective fields. However, the government of Bangladesh should take the necessary and strategic steps to make use of the progress of AI and use this technology to help in the development of society. Everyone must work tirelessly to continue the trend of outstanding progress of Bangladesh and to accelerate the development of industry in this country. Adequate progress needs to be made to improve the proper use of AI at all levels of society. Currently, Bangladesh is not able or ready to use all applications of AI. To a lesser extent, various uses of Bangladesh AI technology have already started. By 2021, the present government wants to implement the National Strategy Vision-2021 where building Digital Bangladesh is one of the priorities. We present a comprehensive picture of AI technology, its potential, scope, and impact. It is clear that AI is becoming an integral part of the global program due to its wide potential. It is certain that AI technology will have an impact on social, economic as well as political planning, and since it is directly linked to employment, it will generate some controversy. Ensuring proper use of this technology will lead to massive development in society and will increase the quality of life of the people. However, to ensure the proper use and benefits of artificial intelligence technology, there must be a specific guideline. If we can ensure good governance in the implementation of AI technology strategy from the beginning, we can accelerate the development of Bangladesh and establish it as a developed country in the court of the world.

References

1. Absar MMN, Amran A, Nejati M (2014) Human capital reporting: evidences from the banking sector of Bangladesh. Int J Learn Intell Cap 11(3):244–258. https://doi.org/10.1504/IJLIC.2014.063899.Accessed15Jan2021
2. Adadi A, Berrada M (2018) Peeking inside the black-box: a survey on explainable artificial intelligence (XAI). IEEE Access 6(9):52138–52160. https://doi.org/10.1109/ACCESS.2018.287%200052.Accessed31July2020
3. Abramowski T (2013) Application of artificial intelligence methods to preliminary design of ships and ship performance optimization. Naval Eng J 3(1):98–98. https://www.researchgate.net/publication/259361068_Application_of_Artificial_Intelligence_Methods_to_Preliminary_Design_of_Ships_and_Ship_Performance_Optimization. Accessed 1 Mar 2021
4. Bangladesh Govt. Report: National Strategy for Artificial Intelligence Bangladesh (2019a). National Strategy for Artificial Intellgence - Bangladesh.pdf. Accessed 23 June 2020
5. Bangladesh Govt. Education Report: Education Performance Report (2019b). https://www.dpe.gov.bd/sites/default/files/files/dpe.portal.gov.bd/publications/07c9f4cb_c14a_49da_9201_7774c2aeef84/Final%20ASPR%202019.pdf. Accessed 24 Aug 2020
6. Choudhury MI, Chowdhury SA, Mahdi AM, Rahaman S (2020) Human resource management practices in bangladesh : a review paper on selective HRM functions. J Soc Sci Educ Human 1(2):43–49. https://www.sciworldpub.com/journal/JSS%20EH. Accessed 22 Feb 2021
7. Deowan SA (2020) Artificial intelligence: Bangladesh Perspective. Bus Stand. https://www.tbsnews.net/tech/artificial-intelligence-bangladesh-perspective-44017. Accessed 25 Apr 2021

8. Fersht P, Gupta S, Pillala A (2019) Bangladesh emerges as a distinctive digital hub for emerging technologies. HFS Res 1(1):1–19. http://lict.gov.bd/uploads/file/strategic/strategic_5ca49063b6714.pdf. Accessed 12 Apr 2021

9. Garg V, Srivastav S, Gupta A (2018) Application of artificial intelligence for sustaining green human resource management. In: 2018 international conference on automation and computational engineering, ICACE 2018, vol 1, no 8, pp 113–116. https://doi.org/10.1109/ICACE.2018.86869%208. Accessed 25 Feb 2021

10. Hasan MJ (2019) Artificial intelligence: benefits and limitations in Bangladesh. City Univ J 2(1):75–84. https://www.researchgate.net/publication/332548432_Artificial_Intelligence_Benefits_and_Limitations_in_Bangladesh. Accessed 12 Dec 2020

11. Hossain MS, Ulfy MA, Ali I, Karim MW (2021) Challenges in adopting artificial intelligence (AI) in HRM practices: a study on Bangladesh perspective. Int Fellowsh J Interdiscip Res 1(1):66–73. https://www.researchgate.net/publication/348916775_Challenges_in_Adopting_Artificial_Intelligence_AI_in_HRM_Practices_A_study_on_Bangladesh_Perspective. Accessed 1 Mar 2021

12. Islam MM, Khondoker M, Rahman CM (2001) Application of artificial intelligence techniques in automatic hull form generation. J Ocean Eng 28(12):1531–1544. https://www.sciencedirect.com/science/article/abs/pii/S0029801801000208. Accessed 10 Apr 2021

13. Li Z, Weimin C (2020) Key technology of artificial intelligence in hull form intelligent optimization. In: ICMAI proceedings of the 5th international conference on mathematics and artificial intelligence. https://doi.org/10.1145/3395260.3395296. Accessed 30 Mar 2021

14. Mathur S (2019) Artificial intelligence: redesigning human resource management, functions and practices. In: Human resource: people, process and technology. https://www.researchgate.net/publication%20/338448468. Accessed 31 Dec 2020

15. Singer PW, Freidman A (2014) Cybersecurity and cyberwar: what everyone needs to Know. Oxford University Press, Oxford

16. Williams B (2014) Cyberspace: what is it, where is it and who cares?. http://www.armedforcesjournal.com/cyberspace-what-is-it-where-is-it-and-who-cares/. Accessed 15 Jan 2021

Combating the Cyber-Security Kill Chain: Moving to a Proactive Security Model

Jim Seaman

Abstract A former boss of mine (Peter Drissell (https://www.linkedin.com/in/peter-drissell-b917896/) (Commandant General RAF Regiment Air Officer Royal Air Force Police) once delivered a presentation at a University lecture, which I had been attending. Here he made the following statement:

> Many business leaders regard Security as being very expensive and virtually invisible. That is until it goes wrong, when it becomes very visible and considerably more expensive!

Ever since hearing this statement, I have sought to change this view. Having a proactive, asset and risk focused approach that is aligned with the business mission statements/objectives has a significant impact on changing the business leaders' perspectives. This chapter seeks to explain how you can start to reduce the opportunities for the cyber-attackers, through a more targeted and prioritized approach. Many organizations are feeling a sense of Cyber-security fatigue and often sensing that the cyber-criminals have got the upper hand and that this is a battle that they are losing, frequently believing that they are 'Boiling the Ocean'. If a business fails to identify and categorize their assets, they will not be able to truly appreciate the value of their most important company assets, and their importance to the business. Consequently, when it comes to carrying out the risk assessments, it can often feel like this is based upon a premonition or a hunch. Additionally, when it comes to applying appropriate mitigation controls, this can be extremely difficult to show proportionality and a return on investment.

Keywords Asset management · Risk management · Cyber-security · Business stakeholders · Mission statements

J. Seaman (✉)
IS Centurion Consulting Ltd., London, UK
e-mail: contact@iscenturion.com

© The Author(s), under exclusive license to Springer Nature Switzerland AG 2021
R. Montasari and H. Jahankhani (eds.), *Artificial Intelligence in Cyber Security: Impact and Implications*, Advanced Sciences and Technologies for Security Applications,
https://doi.org/10.1007/978-3-030-88040-8_5

121

1 Introduction

We are seeing a considerable increase in the number of reported cyber-attacks and organizations that are suffering data breaches. Some of these victims, of the cyber-criminals' activities are not restricted to the small to medium sized businesses and include some large, well-respected, enterprises. This became especially true during the 2020 pandemic, when we observed numerous businesses having to rapidly adjust to new disparate, remote-working business operating models.

However, in most instances it is not the case that the cyber-criminals have started to become increasingly skilled, are using highly sophisticated tactics, techniques & protocols (TTPs) or are employing the very latest tools & technologies.

On the contrary, many of the cyber-criminals are still using some well-established practices to gain unauthorized entry to your corporate environment and valuable assets. Most of today's cyber-criminals are opportunist attackers, seeking to identify the easiest of targets that they can exploit. Although, it is worth noting that there are still those highly organized and well-funded State Sponsored groups, who will have their sights set on undermining the defenses of the better defended high-profile organizations.

Wherever your business resides on the sliding scale of value, there is an enemy out there seeking to target your business and to gain a return on investment (ROI) from your valued assets.

As already mentioned, today's cyber-criminals come in various guises and having varying levels of competence and resources. However, common to all these threat actors are the following three drivers (as depicted in Fig. 1):

Consequently, it is essential that all businesses *(whether large or small)* understand:

– **What assets are important to the continued business operations?**

   Which assets are connected to *(or could impact)* the business critical/important assets?

– **What are the perceived threats to these assets?**
– **What vulnerabilities are associated with these assets?**
– **If these business critical/important assets become compromised, how much impact will this have on the business?**

To enable this, you need to gain a better understanding of your potential attackers, their 'modus operandi' and how to build a more proactive approach, so that you are more capable at identifying and quickly responding to the stages within the Cybersecurity Kill Chain.

In this chapter, I will describe the foundations of a proactive model, through the application of the **ARMED** acronym:

• **ARMED** (**A**sset & **R**isk **M**anagement for **E**nhanced **D**efense).

1. Inquisitiveness.
They want to know what is within your environments that might be beneficial to them.

2. Challenge
Your attacker's get a kick out of coming up with clever or ingenious ways of circumventing any defenses.

3. Reward
Although most cyber-attacks are seeking a return on their effort:
- Monetarizing your valuable assets for their profit.
- Those who will happily attack your business only for the kudos. However, often these treat actors can be the most damaging, where they need to exfiltrate your valuable assets or leave a visible mark (Graffiti) on your assets:
Evidence as to their daring activities.

Fig. 1 Cyber attacker drivers

2 What is the Cyber Kill Chain?

This term was adapted from the military term 'Kill Chain', which was used to describe the stages that an enemy might go through when planning and launching an attack against military establishments and assets.

It became a common place understanding that if you understood the stages that your enemy might go through, then you would be better placed to ensure that your defenses remain effective to quickly identify and respond to any hostile activities. As a result, the defenses would be better placed to contain and limit the potential impact/damage that such attacks could yield.

Throughout military history, there are numerous references to differing variants of the 'Kill Chain' model but essentially all identify that any attack goes through various stages:

F2T2EA [1]

- Find
- Fix
- Track
- Target
- Engage
- Assess.

Later, following the popularity of the internet and the growing number of cyber-attacks, Lockheed Martin adapted the military concept to assist in the defense of cyber-space. They termed this:

The Cyber Kill Chain [2]

Basically, prefixing the military term with the word '**Cyber** [3]':

"word-forming element, ultimately from cybernetics (q.v.). It enjoyed explosive use with the rise of the internet early 1990s. One researcher (Nagel) counted 104 words formed from it by 1994. Cyberpunk (by 1986) and cyberspace (1982) were among the earliest. The OED 2nd edition (1989) has only cybernetics and its related forms, and cybernation "theory, practice, or condition of control by machines" (1962)".

In essence, the Cyber Kill Chain model breaks up an attack into the following 7 distinct steps (as depicted in Fig. 2):

Starting with surveillance, the attackers are looking for opportunities they can use to gain an unauthorized persistent presence within their target's environment. They

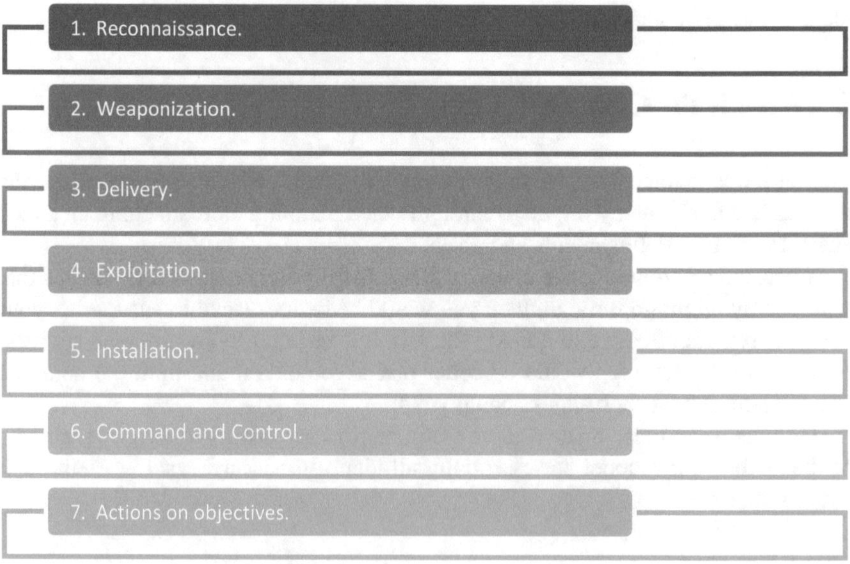

Fig. 2 Cyber kill chain

Fig. 3 Cognitive attack loop

seek to do this as clandestinely as they can, so that they can achieve the Command and Control that they require to exfiltrate the data or launch the damaging attack.

Much the same as the military version of the term, subsequently, there have been several differing variants of this term:

Cognitive Attack Loop [4] (as depicted in Fig. 3).

Mitre Attack Framework [5] (as depicted in Fig. 4).

All these models clearly show the fact that the attackers want to understand is what assets you have, where they reside, to categorize the value of these assets and to understand the layout of your environment.

Long before most attackers 'press the launch button', will have carried out extensive observations, planning and mapping of their target environment and often having several viable options to choose from.

Most today's cyber-criminals are in it for the long haul and will patiently play the waiting game, gently tip toeing their way around their target's infrastructure, often trying not to create too much noise, or testing to see if anyone reacts to a tripped alarm or seeing whether they can cause a distraction to evade being detected.

Unified Kill Chain [6].

Building upon the concepts of all the previous versions of the Cyber Kill Chain, the Unified Kill Chain details 18 distinct stages of an attack (as depicted in Fig. 5).

These distinct stages are then aligned with 4 defined phases of an attack path (as depicted in Fig. 6).

However, it is worth noting that in many cyber-attacks, at the point that the organization has detected the presence of an unauthorized entity, they have already gained access (in the **Step 4** phase).

At this point, it is often too late, as this noise being created is the attackers exfiltrating the data and themselves from their victim's environment.

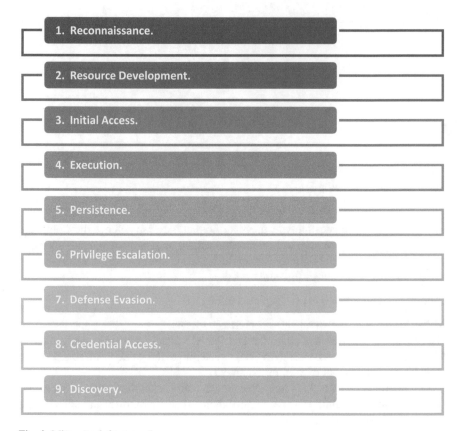

Fig. 4 Mitre attack framework

They've grabbed their 'swag' and are hurriedly making their way towards their preplanned exit!

Consequently, for an effective and proactive defensive model you need to ensure that your defensive efforts can distinguish the **ABNORMAL** activities from the NORMAL activities that you would expect from business-as-usual (BAU) operations and be able to quickly recognize the steps from the Cyber Kill Chain, occurring inside your business environment.

3   Are You ARMED and Ready?

An effective and pro-active model needs to apply some of the tactics seen to be employed by your attackers and this starts by fully understanding your environment and knowing what assets present the greatest importance to the business and to have a comprehensive understanding of the potential risks to your business.

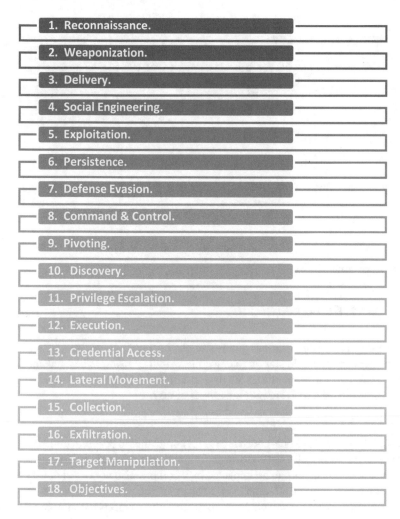

1. Reconnaissance.
2. Weaponization.
3. Delivery.
4. Social Engineering.
5. Exploitation.
6. Persistence.
7. Defense Evasion.
8. Command & Control.
9. Pivoting.
10. Discovery.
11. Privilege Escalation.
12. Execution.
13. Credential Access.
14. Lateral Movement.
15. Collection.
16. Exfiltration.
17. Target Manipulation.
18. Objectives.

Fig. 5 Unified kill chain

Consequently, for a pro-active approach before considering anything else, you need to understand and manage your assets and risks.

We are learning that the cyber-criminals are starting to embrace machine learning and artificial technologies, to assist them with their cyber-attacks [7], so should we be considering the same with our asset and risk management endeavors?

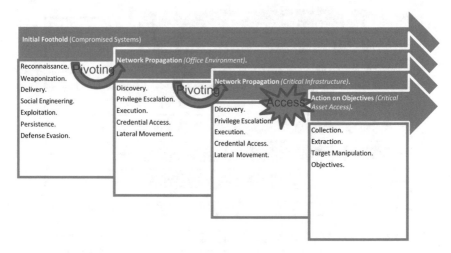

Fig. 6 Attack path

3.1 Asset Management

Frequently, many organizations jump straight into compliance modes and forget that their interests should start with engaging with the business' key stakeholders to identify what business operations are the most important to them and to then identify what assets are critical to support these operations?

What is an asset?

In a digital and technological age, it is easy for companies to become 'tunnel visioned' into centering on the IT assets. However, it is extremely important to remember that most business processes can be impacted by a compromise of critical non-IT assets.

For instance, look at the impact on businesses (caused by the pandemic) when their personnel were prevented from going into their workplace and they had not established remote working in their business continuity planning.

Or, that organization that has a small number of IT support staff and, as a result, have allowed themselves to have sole employees that are dedicated to delivering specialist operations which only they understand and know how to do. The next thing they know, one of the single points of failure (SPoF) gets seriously ill and is admitted to hospital.

- **Suddenly their SPoF fails and they have no back up**.

Consequently, when defining what an asset is, it is important to ensure that you think far broader than it being just your IT assets.

NIST Asset Definitions [8]

- *"A major application, general support system, high impact program, physical plant, mission critical system, personnel, equipment, or a logically related group of systems"*.
- *"Anything that has value to an organization, including, but not limited to, another organization, person, computing device, information technology (IT) system, IT network, IT circuit, software (both an installed instance and a physical instance), virtual computing platform (common in cloud and virtualized computing), and related hardware (e.g., locks, cabinets, keyboards)"*.
- *"An item of value to achievement of organizational mission/business objectives.*

  Note 1: Assets have interrelated characteristics that include value, criticality, and the degree to which they are relied upon to achieve organizational mission/business objectives. From these characteristics, appropriate protections are to be engineered into solutions employed by the organization. Note 2: An asset may be tangible (e.g., physical item such as hardware, software, firmware, computing platform, network device, or other technology components) or intangible (e.g., information, data, trademark, copyright, patent, intellectual property, image, or reputation)".

When considering your assets, you need to look at from a business perspective and identify, and categorize your company operations/processes, based upon their importance to your business mission.

Once you have a priority list of your business operations/processes, you can then start to identify and categorize the assets that support these processes. When considering how to categorize your assets, you should think along the lines of:

- How important are they in support of the continued operation of valuable business processes?
- Are they public facing?
- Are they subject to legal and regulatory obligations (e.g. Financial/Personal data processing).
- Do you have a manual or automated process for detecting those business assets (e.g. IT systems, business applications) that are processing, storing, or transmitting your sensitive data assets?

  – How easily and quickly are you able to locate your sensitive data stores?

- Are they connected to (or able to impact) higher value assets?

For example,

If you have identified a Contact/Call Center as being a valuable and essential part of your business, you will need to break this down into its component parts (as depicted in Fig. 7):

Each of these assets have an importance, in support of the Contact/Call Center operations. However, some are more important than the others and you should

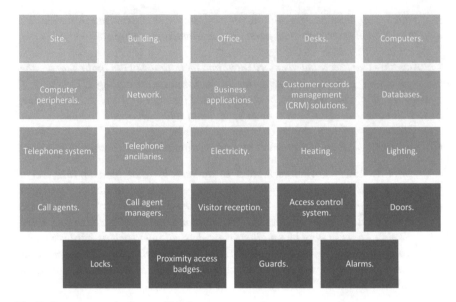

Fig. 7 Components of a Contact/Call Center

consider maintaining a business asset register, third party supplier register and config-uration management database (CMDB), to accurately reflect the assets each valued business operation/process requires.

Additionally, you should also consider creating supporting diagrams to show how the asset dependencies and connections. For example, with a valuable IT asset, connected to the corporate network, you should maintain an accurate network diagram, an example showing the network diagram, for an automatic access control system (AACS), is at Fig. 8. With this, you are better able to easily appreciate and understand the supporting topology, and connectivity.

Where these IT assets have been identified as being essential to the storage, processing or transmission of sensitive business or personal data, you should consider creating and maintaining accurate data flow diagrams (DFD).

Through effective asset management, you should be able to quickly identify any unauthorized, rogue, or dangerous assets that may be attempting to endanger your valued assets or undermine your defensive efforts.

For example.

- A rogue wifi device that is attempting to bypass your firewall.
- A poorly configured unauthorized device, connecting to your corporate network.
- An unauthorized employee attempting to access a restricted area.
- An untethered network connection by a third-party supplier.
- Do you understand the extent of your public-facing digital footprint?

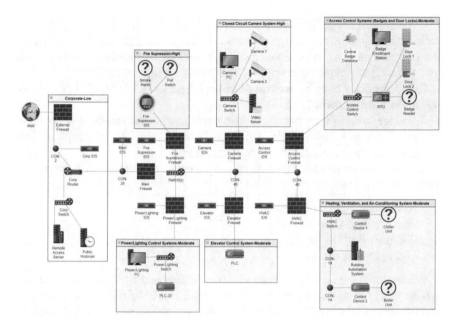

Fig. 8 AACS Network Diagram

- *E.g. Monitoring vulnerabilities through a platform such as Security Scorecard* [9].

- Can you detect a new public-facing IP, associated with your digital footprint domains?
- Do you understand your different scopes, requiring differing levels of protection?

Where there is electronic monitoring of your assets and their behaviors, automation through Machine-Learning or Artificial Intelligence can really help to enhance your asset management capabilities.

As an example.

- How quickly can you detect someone physically accessing a restricted area, outside their normal working hours, or repeatedly mis-entering their physical access personal identity number (PIN)?

  - *Is this a NORMAL or an **ABNORMAL**event?*
  - *Could this be a malicious, accidental, or genuine action?*

Both people and business operations tend to become creatures of habit and set routines. Consequently, where you can employ technology to immediately notify you of any activities that are outside your NORMAL expected parameters would help enhance your ability to effectively respond to the presence of the **ABNORMAL**.

If you are still struggling to understand what you have in your estate and their criticality or are juggling this through several manual processes, you should consider the benefits that technological solutions can bring to enhance your IT asset management [10], enterprise asset management [11] and access management [12].

Cyber Security & Infrastructure Security Agency (CISA) Cyber Resilience [13] **Asset Management** [14].

If you are looking for a comprehensive guide to asset management, CISA have produced a comprehensive guide, which has been developed to help organizations to establish an effective asset management process.

This guide provides an approach that is common to many asset management standards and guidelines:

1. **Planning for asset management.**
2. **Identifying the assets.**
3. **Documenting the assets.**
4. **Managing the assets.**

Readers of this guide can gain an improved understanding about what an effective the asset management process should involve and to help promote a common understanding of the need for an effective asset management process, including such things as:

- Identifying and describing key practices for asset management.
- Providing examples and guidance to organizations wishing to implement these practices.

3.2 Risk Management

It is only once you have established a proactive approach to understanding the assets within your business environment, can you hope to start to understand what is needed to proportionately safeguard these assets—based upon their perceived value to your business.

All too often, many businesses make the mistake of trying to make everything secure and, as a result, some of the higher-risk (more impactful) assets get overlooked. The reason for this is that there is a distinct difference between the terms 'Risk' and 'Security'.

Risk

"1660s, risque, from French risque (16c.), from Italian risco, riscio (modern rischio), from riscare "run into danger," of uncertain origin. The Englished spelling first recorded 1728. Spanish riesgo and German Risiko are Italian loan-words. With run (v.) from 1660s. Risk aversion is recorded from 1942; risk factor from 1906; risk management from 1963; risk taker from 1892."

Definition [15]

"the possibility of something bad happening".

Security [16]

"mid-15c.,

"condition of being secure," from Latin securitas, from securus "free from care" (see secure). Replacing sikerte (early 15c.), from an earlier borrowing from Latin; earlier in the sense "security" was sikerhede (early 13c.); sikernesse (c. 1200).

Meaning "something which secures" is from 1580s; "safety of a state, person, etc." is from 1941. Legal sense of "property in bonds" is from mid-15c.; that of "document held by a creditor" is from 1680s. Phrase security blanket in figurative sense is attested from 1966, in reference to the crib blanket carried by the character Linus in the "Peanuts" comic strip (1956)".

Definition [17]

"protection of a person, building, organization, or country against threats such as crime or attacks by foreign countries".

Consequently, before you can apply proportionate protection you need to understand the value of the assets, as well as their associated threats, vulnerabilities and if the assets are compromised, the potential impacts.

- This is the essence of risk management.

Virtually every business cannot operate their business completely securely and they must accept that there will be some risks to their valuable assets. However, the perceived risks must be balanced between proportionality and functionality.

For example

If your business has a small amount of extremely sensitive data records, which need to be proportionately protected to ensure their Confidentiality, Integrity and Availability is maintained, the most robust cause of action might be to (as depicted in Fig. 11):

1. **Put the extremely sensitive data in a waterproof and fireproof safe.**
2. **Lock the safe.**
3. **Destroy the key.**
4. **Dig a hole in the ground.**
5. **Put the safe in the hole.**
6. **Fill the safe with concrete.**
7. **Build a property onto of the concrete.**

However, the reality is that today's modern businesses are heavily reliant on their data assets and protecting the data in this way may make it more secure but prevents the data from being used in support of business operations.

Consequently, a proactive program needs to look at this using a risk sliding scale, as depicted in Table 1.

Table 1 Risk sliding scale

UNACCEPTABLE					ACCEPTABLE
- Sensitive files left in the open	Sensitive files left in the open. In a locked office. 30 min rule	Sensitive files left in the open. In a locked office. In a locked cabinet	Sensitive files left in the open. In a locked office. In a locked safe.	Sensitive files left in the open. In a locked office. Secure Room. In a locked safe.	Sensitive files left in the open. In a locked Secure Room. In a locked safe. Periodic out of hours security checks.

- The greater the value, the greater the risk=

  – The greater the number of defenses required needed to safeguard the assets, to within acceptable risk levels.

Let's face it, everyone makes risks decisions every day and this allows you to make and informed decision on the best cause of action that is best for you.

Think about it:

Deciding whether to cross a road (as shown in Fig. 9):

Deciding whether to eat that chili pepper *(or not)* (as shown in Fig. 10).

With these examples in mind, it is extremely rare that someone would choose their mitigation options without understanding the associated risks. However, it is common place to hear businesses that are applying mitigation security controls when they do not fully understand the value of their assets and the risks that are associated with them.

Consequently, the traditional approach to securing business assets is often seen as providing little of no return on any investments and especially if after all the investments made, the business still feel the impact of a valued asset suffering a compromise of its Confidentiality, Integrity or Availability.

Threat Modelling

It is worthwhile understanding the threats that pertain to your valued assets, but what is a threat?

Threat (n.) [18]

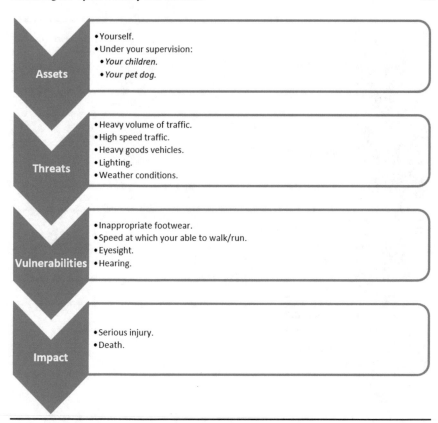

Assets
- Yourself.
- Under your supervision:
  - *Your children.*
  - *Your pet dog.*

Threats
- Heavy volume of traffic.
- High speed traffic.
- Heavy goods vehicles.
- Lighting.
- Weather conditions.

Vulnerabilities
- Inappropriate footwear.
- Speed at which your able to walk/run.
- Eyesight.
- Hearing.

Impact
- Serious injury.
- Death.

Mitigation

Use the pedestrian crossing.
Use a bridge.
Find a more suitable crossing place.

Fig. 9 Road crossing risk assessment

"Old English þreat "crowd, troop," also "oppression, coercion, menace," related to þreotan "to trouble, weary," from Proto-Germanic *thrautam (source also of Dutch verdrieten, German verdrießen "to vex"), from PIE *treud "to push, press squeeze" (source also of Latin trudere "to press, thrust," Old Church Slavonic trudu "oppression," Middle Irish trott "quarrel, conflict," Middle Welsh cythrud "torture, torment, afflict"). Sense of "conditional declaration of hostile intention" was in Old English".

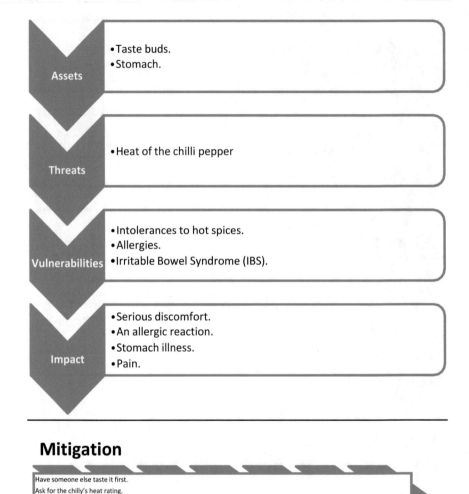

Fig. 10 Chilli eating risk assessment

Threat Definition [19]

"In business analysis, Threats are anything that could cause damage to your organization, venture, or product. This could include anything from other companies (who might intrude on your market), to supply shortages (which might prevent you from manufacturing a product).

Threats are negative, and external. This mean that threats do not benefit your company, but there is nothing you can do to stop them from coming about.

Threats are like opportunities in that you cannot change their frequency, or purposefully bring them about, but you can still choose how to approach them and deal with them".

NIST Threat Definition [20]

"Any circumstance or event with the potential to adversely impact organizational operations (including mission, functions, image, or reputation), organizational assets, or individuals through an information system via unauthorized access, destruction, disclosure, modification of information, and/or denial of service. Also, the potential for a threat-source to successfully exploit a particular information system vulnerability."

To better understand the threats to your valued assets, you need to have a good appreciation of the threats that might be relevant to these assets. Threats can come in both the traditional and non-tradition types. Traditional threats being best described using the military term TESSOC:

- **Traditional Threats**

Terrorism, Espionage, Sabotage, Subversion and Organized Crime (TESSOC) [21].

- **Non-Traditional Threats**

Theft, Natural Disaster (Pandemic, Fire, Flood, Earthquake, etc.), Investigative Journalist, Hacker, Accident, etc.

To understand how and which of these threats might be relevant for your business, threat modelling becomes immensely helpful.

Threat modeling is a method by which you locate and identify your vulnerabilities, identifying objectives, and then design suitable countermeasures to either prevent or mitigate the effects of cyber-attacks against the system.

To create effective threat models, you should be asking yourself the following types of questions:

- **Which assets are the most valuable and need threat models?**
- **What kind of threat model are required?**

  - The answer requires studying network diagrams, data flows, asset inventories/registers, site plans, etc.

    This will help you to create a virtual model of the network you're trying to protect.

- **What are the potential problems/issues?**

  - Here's where you discover the main threats to your valued assets.

- **What actions should be taken to recover from a potentially serious incident?**

  - You've identified some potential problems now; it's time to work out some actionable resolutions.

- **Was it successful?**

  – This step is a follow-up where you conduct a retrospective to monitor the quality, feasibility, planning, and progress.

  Basically, you are looking at your business through the eyes of a potential attacker.

- **What would be valuable to your enemies?**
- **What tactics are your enemies known to employ?**
- **Are you vulnerable to such attacks?**
- **Can you see potential opportunities?**

  There are many threat models [22] that you can apply, in support of your risk management practices, including:

Microsoft's STRIDE model [23].

- *Spoofing.*

  – *Involves illegally accessing and then using another user's authentication information, such as username and password.*

- *Tampering.*

  – *Involves the malicious modification of data. Examples include unauthorized changes made to persistent data, such as that held in a database, and the alteration of data as it flows between two computers over an open network, such as the Internet.*

- *Repudiation.*

  – *Associated with users who deny performing an action without other parties having any way to prove otherwise—for example, a user performs an illegal operation in a system that lacks the ability to trace the prohibited operations. NonRepudiation refers to the ability of a system to counter repudiation threats. For example, a user who purchases an item might have to sign for the item upon receipt. The vendor can then use the signed receipt as evidence that the user did receive the package.*

- *Information Disclosure.*

  – *Involves the exposure of information to individuals who are not supposed to have access to it—for example, the ability of users to read a file that they were not granted access to, or the ability of an intruder to read data in transit between two computers.*

- *Denial of Service.*

  – *Denial of service (DoS) attacks deny service to valid users—for example, by making a Web server temporarily unavailable or unusable. You must protect against certain types of DoS threats simply to improve system availability and reliability.*

- **Elevation of Privilege.**

  - *An unprivileged user gains privileged access and thereby has sufficient access to compromise or destroy the entire system. Elevation of privilege threats include those situations in which an attacker has effectively penetrated all system defenses and become part of the trusted system itself, a dangerous situation indeed".*

Process for Attack Simulation and Threat Analysis (PASTA) [24].

- Define business objectives.
- Define the technical scope of assets and components.
- Application decomposition and identify application controls.
- Threat analysis based on threat intelligence.
- Vulnerability detection.
- Attack enumeration and modeling.
- Risk analysis and development of countermeasures.

OCTAVE [25]

"Operationally critical threat, asset, and vulnerability evaluation (OCTAVE) is an approach to identify, assess, and manage risks to IT assets.

This process identifies the critical components of information security and the threats that could affect their confidentiality, integrity, and availability. This helps them understand what information is at risk and design a protection strategy to reduce or eliminate the risks to IT assets".

Attack Trees [26].

"The tree is a conceptual diagram showing how an asset, or target, could be attacked, consisting of a root node, with leaves and children nodes added in. Child nodes are conditions that must be met to make the direct parent node true. Each node is satisfied only by its direct child nodes.

It also has "AND" and "OR" options, which represent alternative steps taken to achieve these goals".

Vulnerability Modelling

Having identified and gained a better understanding of the threats that pertain to your valued business assets, you then need to gain a better understanding of the vulnerabilities that are associated with these important business assets, which could be exploited by your threat actors.

Vulnerability (n.) [27]

"1767, noun from vulnerable (q.v.)."

Vulnerable (adj.) [28]

c. 1600, from Late Latin vulnerabilis "wounding," from Latin vulnerare "to wound, hurt, injure, maim," from vulnus (genitive vulneris) "wound," perhaps related to vellere "pluck, to tear" (see svelte), or from PIE *wele-nes-, from *wele (2) "to strike, wound" (see Valhalla)".

NIST Vulnerability Definition [29]

"Weakness in a system, system security procedures, internal controls, or implementation that could be exploited or triggered by a threat."

In essence, you are looking to identify any exploitable weaknesses that are associated with those assets that need to be adequately protected. Where you have linked assets, you should consider the implications of any vulnerabilities to connected assets, which could be present an aggregated risk.

Each vulnerability should be assessed to ascertain their potential repercussions on your business operations if an opportunist attacker were to take advantage of this weakness.

All vulnerabilities should be identified and prioritized against the safeguarding of the business operations, and remediation road maps maintained to ensure that any high-risk vulnerabilities do not remain beyond the acceptable time scales.

Impact Modelling

Just because you have identified some threats and vulnerabilities against your valued business assets, this does not mean that this will present an impact to your business operations, which you need to be concerned about.

Impact (v) [30]

"c. 1600, "press closely into something," from Latin impactus, past participle of impingere "to push into, drive into, strike against," from assimilated form of in "into, in, on, upon" (from PIE root *en "in") + pangere "to fix, fasten" (from PIE root *pag "to fasten"). Original sense is preserved in impacted teeth. Sense of "strike forcefully against something" first recorded 1916. Figurative sense of "have a forceful effect on" is from 1935. Related: Impacting".

NIST Impact Definition [31]

"With respect to security, the effect on organizational operations, organizational assets, individuals, other organizations, or the Nation (including the national security interests of the United States) of a loss of confidentiality, integrity, or availability of

information or a system. With respect to privacy, the adverse effects that individuals could experience when an information system processes their PII."

However, without an effective business impact analysis (BIA) [32], you will not be able to make an informed choice on whether you are comfortable with the potential impact, or whether you need to reduce the potential impacts to within acceptable levels.

3.3 Initial Risk Management

Before you can start to think about the application of any mitigation controls, you need to understand the level of risks that are inherent to your valued business assets and to what levels of risks your organization's risk owners are comfortable with. This is termed as identifying the inherent risk and defining the risk appetites:

Initial Risk [33]

"Risk before controls or countermeasures have been applied."

Risk Appetite [34]

"The types and amount of risk, on a broad level, an organization is willing to accept in its pursuit of value."

Understanding these terms and how they are applied/implemented in your organization are essential to ensuring that any mitigation efforts remain proportionate to the expectations of the business. You want to ensure that the business is comfortable with the perceived levels of risk and that you're not applying additional defensive measures than the business needs.

3.4 Risk Management Cycle

There are several risk management frameworks that your organization may choose to baseline against (e.g., NIST Risk Management Framework (RMF) [35]) they all involve several logical steps that should be followed, such as those within the NIST RMF [36], as depicted in Table 2.

However, no matter which specific methodology that a business chooses, most incorporate six key areas of the risk management cycle, as detailed below.

1. **Set strategy**.
2. **Identify risk**.
3. **Prioritize risk**.
4. **Assess mitigation controls**.
5. **Monitor**.

Table 2 NIST RMF

Prepare	Standards	
1. Categorize System	FIPS 199/SP800-60/CUI Registry	SP800-30
2. Select Controls	FIPS 200/SP800-53/SP800-53B	
3. Implement Controls	Multiple NIST Publications (e.g. SP800-34, SP800-61, SP800-128, etc.)	IR 8062
4. Assess Controls	SP800-53A/IR 8011	
5. Authorize System	SP800-37	SP800-160
6. Monitor Controls	SP800-37/SP800-53A/SP800-137/SP800-137A/IR 8212	SP800-18

6. **Measure.**

Key objectives

- The strategy for managing risk is set by the board.
- There are distinct processes both to identify and to prioritize risk.
- The existing controls and monitoring procedures are then assessed.
- There is a process to measure the residual risk position and to monitor progress going forward.

Much like running a business, Risk Management should be regarded as a living process, which needs to be subject to dynamic and ongoing management, ensuring that a team effort is applied to help assure that timely identification, assessment, and remediation of risks is maintained.

3.5 Risk Assessment

When commencing risk assessments, you need to ensure that you start with the categorization of your efforts so that they align with the business' priorities. You don't want to be focusing your efforts on risk assessing a lower value business process area, ahead of higher value parts of the business.

An effective risk assessment will clearly articulate the perceived levels of risk that the assessed business area is facing and, where these are observed to be above the levels with which the risk owners are comfortable with, you are then better placed to select and suggest appropriate mitigation security controls that can be used to reduce these risk levels, to within acceptable parameters. You should select a range of viable courses of action with which the risk owners can choose from and sign off on.

This decision-making process should be documented, to show governance and that the risk owners were presented the opportunity to make an informed choice as to the best courses of action.

For me, the easiest way to remember and present the course of action options is through the 4 Ts of Risk Management [36]:

- Treat.
- Tolerate.
- Terminate.
- Transfer.

Having successfully identified the potential risks, along with a choice of suitable mitigation solutions, the chosen best course of action needs to be implemented and the risk re-assessed to identify the residual risk and ensure that it has been reduced to within acceptable tolerances.

Residual Risk [38]

"The potential for the occurrence of an adverse event after adjusting for the impact of all in-place safeguards."

3.6 Quality Versus Quantity

Frequently a great deal of businesses either pay lip service to risk management or adopt the simpler risk model of Qualitative risk assessments. However, with risk being so integral and important to the safeguarding of your valued business assets and providing visibility of the return on investment for your risk mitigation, it is beneficial to employ both Qualitative and Quantitative risk assessments.

Qualitative Risk

NIST [39] defines a Qualitative assessment as being:

"Use of a set of methods, principles, or rules for assessing risk based on nonnumerical categories or levels".

Qualitative assessment employs an element of subjective judgment to analyze an organization's risk based on non-quantifiable information, where individuals are given a reference scale, based upon the organizations risk tolerances. These reference scales are then employed to help with the forecasting of the potential probability and likelihood of the perceived risks, as depicted in Table 3.

These results are then plotted onto a risk heat map (as depicted in Table 4) helping to visualize and scale the risks. However, often this can feel like a 'finger in the air' assessment and can be difficult to quantify the potential benefits of any risk mitigation measures.

Consequently, for a comprehensive risk management approach qualitative risk has its place for visualizing risk profiles but these needs to be supplemented by quantitative risk assessments, to help quantify the benefits of any risk mitigation efforts.

Quantitative Risk

NIST [40] defines a Quantitative assessment as being:

Table 3 Scales of risk

Likelihood (Probability) Value Scales		
Level		**Description**
Certain	5	*Critical likelihood of occurrence.* *Threat expected monthly.*
Likely	4	*High likelihood of occurrence.* *Threat expected on a quarterly basis.*
Possible	3	*Medium likelihood of occurrence.* *Threat expected on a 6-monthly basis.*
Unlikely	2	*Low likelihood of occurrence.* *Threat is expected annually.*
Improbable	1	*Extremely low likelihood of occurrence.* *Threat is seldom to happen (several years).*
Impact Value Scales = Average Scale of C. I. A.		
Level		**Description**
Critical	5	*Threats have a critical impact/effect on organization's reputation (e.g. Industrial espionage, accidental or intentional leakage of critical security information to external enemies, etc.).*
High	4	*Deliberate threats, any occurrence that has a premeditated intent, for example, include a mal-content, important data leakage or modification by an employee, employee unauthorized shredding of important documents, etc. (Unauthorized access, Social engineering).*
Medium	3	*Accidental threats, any occurrence that doesn't have a premeditated intent, for example, an employee accidentally deleting an important file, failed back-up, etc. (Operational user errors).* *Natural and Environmental threats (e.g. Earthquake, Lightening, High temperature, etc.)*
Low	2	*Natural and Environmental threats (e.g. Earthquake, Lightening, High temperature, etc.)*
Insignificant	1	*Threat source is neither motivated or capable.*

"Use of a set of methods, principles, or rules for assessing risks based on the use of numbers where the meanings and proportionality of values are maintained inside and outside the context of the assessment".

With a quantifiable risk assessment, you can see a monetary value of the potential risks and the reduced costs associated to any mitigation efforts. An example, of a quantitative risk assessment is provided at Table 5 [41].

Your risk analysis engagements can be further enhanced through risk analysis with the PESTLE strategic planning model, as depicted in Table 6 [42].

The advantages of PESTLE analysis [43] is that it can prove to be more cost effective, provides a deeper understanding of your business, awareness of the threats, and the potential methods available to exploit opportunities.

Table 4 Example risk heat matrix

ID	Risk Name				
1	Confidentiality of Information				
2	Integrity of Information				
3	IT service disruptions, due to poor BCP/DR planning – involving critical systems.		High	6	3	1
4	IT service disruptions, due to poor BCP/DR planning – involving non-critical systems.	Likelihood				
5	Fraudulent user activities.		Medium	8	9	2
6	Inadequate change management process.					
7	Security configuration management.					
8	Threat Intelligence and Security Event Monitoring.		Low	4	7	5
9	Vulnerability Management.					
				Low	Medium	High
	Legend			Impact		

Critical	High
Medium	Low

4 Post Risk Assessment

Having established the importance of having effective asset and risk management processes, it is now essential that any risk assessments can be presented to the appropriate risk owners so that they can make an informed choice on which risk treatment options they think are the most suitable for bringing the risk to within a range in which they are comfortable (Risk Appetite) or deciding whether to escalate to a more senior employee who has been delegated a greater scales of risk appetite.

Each risk decision should be documented and recorded, providing a centralized and hierarchical view of your organization's risk profile (e.g. Company risk, Geographical risk, Department risk, etc.), as depicted in Fig. 11 [44].

When presenting the risk treatment options, it is extremely important that it is presented in a manner that the Risk Decision Owner can fully understand and appreciate the risks, so that they are able to make an informed decision as to their preferred course of action (CoA).

Table 5 Quantitative risk assessment

Phishing Risk	Mitigation Measures
The Annualized Loss Exposure (ALE) that results from the estimated probable frequency and probable magnitude of future loss for this scenario.	

	Mitigation Measures
£20.4M (Minimum), £50.4M (Average), £110.5M (Maximum); Loss Exceedance Curve	← Employ Anti-spoofing controls
	← Reduce digital footprint
	← Filter/Block incoming emails
	← Train users to identify & report suspected phishing emails
	← Implement multi-factor authentication (MFA)

Summary of Simulation Results

Primary

	Min	Avg	Max
Loss Events / Year	3	4.16	6
Loss Magnitude	£3.3M	£7.8M	£13.5M

Secondary

	Min	Avg	Max
Loss Events / Year	1	4.03	6
Loss Magnitude	£3.4M	£4.5M	£6.8M
Vulnerability			**99.98%**

For example

1. **Define the Problem.**

   a. Risk
      (1) Description of the consideration/deduction

2. **Evaluation of Factors**

   a. Likelihood
      (1) Threat
         i.   Description of the consideration/deduction
         ii. Description of the task/constraint
      (2) Vulnerabilities
         i.   Description of the consideration/deduction
         ii. Description of the task/constraint
      (3) Impact
         i.   Description of the value
         ii. Description of the output

Table 6 PESTLE risk analysis

Political	*Factors may be altered by the government's influence on a country's infrastructure. This may include tax policy, employment laws, environmental regulations, trade restrictions, tariffs, reform and political stability. Charities may need to consider where a government does not want services or goods to be provided.*
Economic	*Factors include economic growth, interest rates, exchange rates, inflation, wage rates, working hours and cost of living. These factors may have major impacts on how charities operate and make decision.*
Social	*Factors include cultural aspects, health and safety consciousness, population growth rate and various demographics.*
Technological	*Factors include ecological and environmental aspects and available products and services. Charities may need to innovate, having considered the compatibility with their own technologies and whether they are transferable internationally.*
Legal	*Factors include any law which may impact on the charities' operations, including NGO regulation and criminal and terrorist legislation which will differ from country to country.*
Environmental	*Factors include an awareness of climate change or seasonal or terrain variations which may affect charities' service delivery methods.*

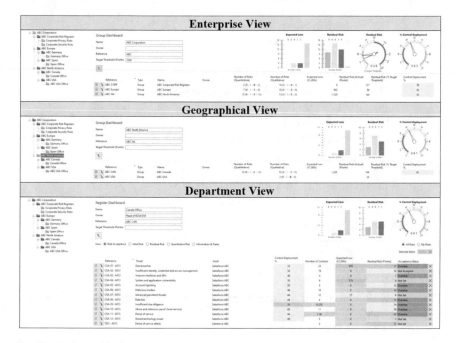

Fig. 11 Business risk views

3. **Courses of Action (CoA)**
(1) Intent

   i.   Description of the consideration/deduction
   ii. Description of the task/constraint

(2) Commonalities between COAs
(3) CoA 1

   i.   Description of the consideration/deduction
        (a) Advantages
        (b) Disadvantages
   ii. Description of the task/constraint

(4) CoA 2

   i.   Description of the consideration/deduction
        (a) Advantages
        (b) Disadvantages
   ii. Description of the task/constraint

(5) CoA 3

   i.   Description of the consideration/deduction
        (a) Advantages
        (b) Disadvantages
   ii. Description of the task/constraint

(6) CoA 4

   i.   Description of the consideration/deduction
        (a) Advantages
        (b) Disadvantages
   ii. Description of the task/constraint

(7) Security Adviser Recommended CoA

   i.   Description of the consideration/deduction
   ii. Description of the task/constraint

4. **Risk Owner's Decision**

   a. Selection of CoA
       (1) Preferred CoA
   b. Implementation plan
       (1) Description of how the mitigation controls are to be implemented.
   c. Date of next review
   d. Appointment of Risk Manager.

You can see that this shows a chain of ownership and accountability for the risk decision process, as well as the management of the risk. However, you will also see that in this example, I have provided 4 CoAs. The reason being that if I were to provide an example with just 3 CoAs, you may fall into the common trap of providing the following CoAs:

1. **The 'Luxury' CoA**

   – A choice that is comprehensive but completely unproportionate to the perceived value of the asset needing protection.

2. **The 'Practical' CoA**

   – The preferred choice of the person submitting the risk treatment options.

3. **The 'Poor' CoA**

   – A completely inadequate or ineffective risk treatment option.

Remember, that all the risk treatment options need to be viable options, so that the risk decisions are made against plausible CoAs and not just tunneling the risk making process into a single option, and the risk owner may even decide to choose not to follow the security advisor's recommendation.

Having selected the preferred CoA, you should carry out a supplementary risk assessment and document the results. The best way to enhance your proactive security model, is to centralize everything, using an appropriate risk management platform. However, many organizations will struggle on trying to achieve this using the extremely time-consuming and inefficient use of several different spreadsheets and then trying to collate this and provide the various risk views/dashboards that the organization requires.

Consequently, as the proactive security model revolves around effective asset and risk management, you should really be seriously considering putting a suitable software platform at the heart of your proactive security strategy, so that you can instantly gain an appreciation of your asset and risk statuses (as depicted in Fig. 12 [45]).

Cyber Security & Infrastructure Security Agency (CISA) Cyber Resilience Risk Management [46]

If you are seeking to learn more about the Risk Management process, CISA have created an extremely useful and informative guide to help you establish an effective risk management process for your business.

Readers will gain a common understanding of the risk management process and be better placed to identify and describe the key risk management practices, including:

- Identify risks to which the organization is exposed.
- Analyze risks and determine appropriate risk disposition.
- Control risks to reduce probability of occurrence and/or minimize impact.
- Monitor risks and responses to risks and improve the organization's capabilities for managing current and future risks.

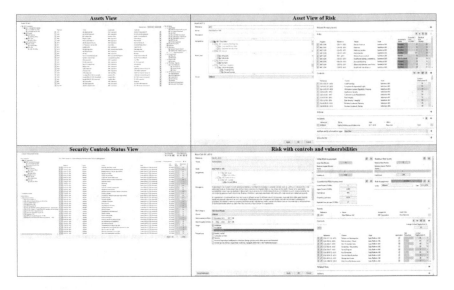

Fig. 12 Assets overview

5 Selection of Mitigation Security Controls

Having identified your assets and their associated initial risks, you will then need to select appropriate and proportionate security controls that can be implemented and maintained. When selecting your risk mitigation controls, you can develop your own or select from the numerous industry security resources and controls that are available to business, e.g.,

- **Common Controls** [47].
- **NIST SP800-53 r5: Security and Privacy Controls for Information Systems and Organizations** [48].
- **DoD Security Technical Implementation Guides (STIGs) and Security Requirements Guides (SRGs)** [49].
- **Community Gold Standard** [50].
- **CSA Cloud Controls Matrix (CCM)** [51].
- **ISO/IEC 27,001/2: Information technology—Security techniques—Information security management systems—Requirements** [52].
- **ISO/IEC 27,701:2019: Security techniques—Extension to ISO/IEC 27,001 and ISO/IEC 27,002 for privacy information management—Requirements and guidelines** [53].
- **ISO/IEC CD 27,402: Cybersecurity—IoT security and privacy—Device baseline requirements** [54].
- **PCI DSS** [55].
- **CIS 20 CSCs** [56].
- **ISACA COBIT 2019** [57].

- **Canadian Baseline cyber security controls for small and medium organizations** [58].
- **UK Government Cyber Essentials** [59]
- **CERT Cyber Resilience Maturity Model** [60].
- **OWASP Application Security Verification Standard (ASVS)** [61].
- **OWASP Mobile Application Security Verification Standard (MASVS)** [62].

5.1 Establishing a Security Baseline

Now, you've established that you have valuable business assets and operations which, if compromised, could impact your organization and that the risks above the levels that you are comfortable with.

Next, you now need to identify the suite of suitable security controls that will provide you with a baseline, with which you are comfortable.

Data Privacy Baseline: Call Center Operations

Your organization has a business unit that provides Call Center operations, and, after the risk assessment, you are far from being happy at the level of risks that this brings to your company. Consequently, you decide to identify some suitable CoAs:

- Terminate.

  Decide that the risks outweigh the business benefits and decide to disband the Call Center operations, in favor of an alternative business operation.
- Tolerate.

  Do nothing and hope that these data processing operations are not compromised.
- Transfer.

  Consider using a third-party service provider to obfuscate the sensitive data, by them converting it into other data formats (e.g. Tokenized Data [63], Dual tone multi-frequency (DTMF) [64], etc.).

  – *No longer will your Call Centre IT systems and personnel have a need to interact with any sensitive data assets. However, although the responsibilities have been transferred to the third-party service provider, you still will be accountable for managing that third-party relationship and to ensure that this supplier continues to operate securely, and in line with your expectations.*
  – *The questions here are:*

    How great is the risk?
    How much cost and effort will it take to reduce the risks to a comfortable level, by implementing inhouse security controls?
    Can the use of outsourced operations enhance the business operations or customer experience?

What is the ROI for using the third-party service?

– Treat.

Reduce the risk by applying the controls from the Common Controls Privacy protection for information and data domain (personal data) and PCI DSS (cardholder data).

By all means consider the added value and assurance of having a cherished business operation or process independently certified against an industry security standard (e.g. ISO/IEC 27001:2013 certification, by an accredited auditor). However, when creating your proactive security model, you should not limit yourself to a single security or compliance framework and ensure that you have a flexible approach that selects the most appropriate security controls *(perhaps chosen from various industry security standards/frameworks)* to help ensure that the perceived risks to your valued business assets remain within your acceptable tolerances.

6   Conclusion

Understanding your assets and the risks associated with them should be your number one priority and any strategy should focus on this. Any mitigation security controls should provide a quantifiable benefit in reducing the risks to your organization and not be purely concentrated on an individual compliance or security controls framework.

With asset and risk management being at the heart of an effective proactive security model, you should ensure that your business key stakeholders are provided with a hierarchical and centralized view of the business' valued assets, their associated risks and mitigation security controls status.

The adoption of this proactive approach requires a team effort but will help to harmonize the approach and provide additional visibility and assurance to the key stakeholders, whilst maintaining the risks to within acceptable levels of tolerance. As a result of this approach, the ROI will become more apparent and will improve the understanding that the business benefits that your security strategy is providing.

References

1. www.militarydictionary.org (n.d.) F2T2EA acronym definition—MilitaryDictionary. https://www.militarydictionary.org/acronym/m/f2t2ea. Accessed 10 Feb 2021
2. Lockheed Martin (2019) Cyber Kill Chain®. [online] Lockheed Martin. https://www.lockheedmartin.com/en-us/capabilities/cyber/cyber-kill-chain.html.
3. www.etymonline.com (n.d.) cyber | search online etymology dictionary. https://www.etymonline.com/search?q=cyber. Accessed 10 Feb 2021
4. https://www.carbonblack.com/blog/introducing-the-cognitive-attack-loop-and-its-3-phases/
5. https://attack.mitre.org/

6. Pols P (2017) The unified kill chain designing a unified kill chain for analyzing, com-paring and defending against cyber attacks. https://www.csacademy.nl/images/scripties/2018/Paul_P ols_-_The_Unified_Kill_Chain_1.pdf.
7. www.trendmicro.com (n.d.) Exploiting AI: how cybercriminals misuse and abuse AI and ML—Security news. https://www.trendmicro.com/vinfo/hk/security/news/cybercrime-and-digital-threats/exploiting-ai-how-cybercriminals-misuse-abuse-ai-and-ml. Accessed 10 Feb 2021
8. Editor CC (n.d.) asset(s)—Glossary | CSRC. [online] csrc.nist.gov. https://csrc.nist.gov/glo ssary/term/asset.
9. partners.securityscorecard.com (n.d.) Cyber rescue alliance—Member | securityScore-card partner portal partner directory. https://partners.securityscorecard.com/english/directory/par tner/462331/cyber-rescue-alliance. Accessed 10 Feb 2021
10. Stone M, Irrechukwu C, Perper H, Wynne D, Kauffman L (2018) IT asset management: financial services. https://csrc.nist.gov/publications/detail/sp/1800-5/final.
11. Inc G (n.d.) Enterprise asset management (EAM) software reviews 2021 | gartner peer insights. [online] Gartner. https://www.gartner.com/reviews/market/enterprise-asset-manage ment-software. Accessed 10 Feb 2021
12. Inc G (n.d.) Network access control (NAC) solutions reviews 2021 | gartner peer In-sights. [online] Gartner. https://www.gartner.com/reviews/market/network-access-control. Accessed 10 Feb 2021
13. us-cert.cisa.gov (n.d.) Assessments: cyber resilience review (CRR) | CISA. https://us-cert.cisa.gov/resources/assessments
14. CRR Supplemental Resource Guide Asset Management (n.d.) https://us-cert.cisa.gov/sites/def ault/files/c3vp/crr_resources_guides/CRR_Resource_Guide-AM.pdf
15. Cambridge.org (2019) RISK | meaning in the Cambridge English Dictionary. https://dictionary.cambridge.org/dictionary/english/risk
16. www.etymonline.com (n.d.) Security | origin and meaning of security by Online Etymology Dictionary. https://www.etymonline.com/word/security#etymonline_v_30368. Accessed 10 Feb 2021
17. Cambridge.org (2019) SECURITY | meaning in the Cambridge English Dictionary.https://dic tionary.cambridge.org/dictionary/english/security
18. www.etymonline.com (n.d.) Threat | search online Etymology Dictionary.https://www.etymon line.com/search?q=threat&ref=searchbar_searchhint. Accessed 10 Feb 2021
19. Frue K (2019) PESTLE analysis—Business and SWOT analysis. [online] PESTLE analysis. https://pestleanalysis.com
20. Nist.gov (2015) Threat—Glossary | CSRC. https://csrc.nist.gov/glossary/term/threat
21. Royal Navy MOD UK (2017) CHAPTER 29 ESTABLISHMENT/UNIT SECURITY OFFICER. Duties of the Establishment/Unit Security Officer. Accessed 10 Feb 2021
22. Exabeam (2020) 6 threat modeling methodologies: prioritize & mitigate threats. https://www.exabeam.com/information-security/threat-modeling. Accessed 10 Feb 2021
23. jegeib (n.d.) Threats—Microsoft threat modeling tool—Azure. [online] docs.microsoft.com. https://docs.microsoft.com/en-us/azure/security/develop/threat-modeling-tool-threats
24. Reliable Cyber Solutions (2020) PASTA threat modeling method: all you need to know—RCyberSolutions.com.https://www.rcybersolutions.com/pasta-threat-modeling-method-all-you-need-to-know. Accessed 10 Feb 2021
25. EC-Council (n.d.) Threat modeling | importance of threat modeling. https://www.eccouncil.org/threat-modeling. Accessed 10 Feb 2021
26. Simplilearn.com (2020) What is threat modeling: process and methodologies. https://www.sim plilearn.com/what-is-threat-modeling-article
27. www.etymonline.com (n.d.) vulnerability | search online etymology dictionary. https://www.etymonline.com/search?q=vulnerability&ref=searchbar_searchhint. Accessed 10 Feb 2021
28. www.etymonline.com (n.d.) vulnerable | origin and meaning of vulnerable by online etymology dictionary. https://www.etymonline.com/word/vulnerable. Accessed 10 Feb 2021
29. Nist.gov (2015) vulnerability—Glossary | CSRC. https://csrc.nist.gov/glossary/term/vulnerabi lity

30. www.etymonline.com (n.d.) impact | origin and meaning of impact by online etymology dictionary. https://www.etymonline.com/word/impact#etymonline_v_1545. Accessed 10 Feb 2021

31. Editor CC (n.d.) Impact—Glossary | CSRC. [online] csrc.nist.gov.https://csrc.nist.gov/glo ssary/term/impact. Accessed 10 Feb 2021

32. Excel TMP (2016) Business impact analysis template excel. https://exceltmp.com/business-impact-analysis-template-excel. Accessed 10 Feb 2021

33. IADC Lexicon (2017) Definition of initial risk. https://www.iadclexicon.org/initial-risk. Accessed 10 Feb 2021

34. Editor CC (n.d.) Risk appetite—Glossary | CSRC. [online] csrc.nist.gov. https://csrc.nist.gov/ glossary/term/Risk_Appetite. Accessed 10 Feb 2021

35. nicole.keller@nist.gov (2020) Risk management framework. [online] NIST. https://www.nist. gov/cyberframework/risk-management-framework

36. Blank R, Gallagher P (2012) Guide for conducting risk assessments NIST special publication 800–30 Revision 1 JOINT TASK FORCE TRANSFORMATION INITIATIVE. https://nvl pubs.nist.gov/nistpubs/legacy/sp/nistspecialpublication800-30r1.pdf

37. Giles S (2012) Managing fraud risk : a practical guide for directors and managers. Wiley, Chichester, West Sussex

38. Editor CC (n.d.) Residual risk—Glossary | CSRC. [online] csrc.nist.gov. https://csrc.nist.gov/ glossary/term/residual_risk. Accessed 10 Feb 2021

39. Editor CC (n.d.) Qualitative assessment—Glossary | CSRC. [online] csrc.nist.gov. https://csrc. nist.gov/glossary/term/Qualitative_Assessment. Accessed 10 Feb 2021.

40. Editor CC (n.d.) Quantitative assessment—Glossary | CSRC. [online] csrc.nist.gov. https:// csrc.nist.gov/glossary/term/Quantitative_Assessment. Accessed 10 Feb 2021

41. app.fairu.net (n.d.) FAIR-U. https://app.fairu.net. Accessed 10 Feb 2021

42. Tool 3: Risk management (n.d.). https://assets.publishing.service.gov.uk/government/uploads/ system/uploads/attachment_data/file/550691/Tool_3.pdf

43. Bush T (n.d.) 3 tools to include in risk management framework for best results. [online] pestleanalysis.com.https://pestleanalysis.com/risk-management. Accessed 10 Feb 2021

44. Acuity Risk Management (n.d.) STREAM integrated risk management software. https://acu ityrm.com. Accessed 10 Feb 2021

45. Acuity Risk Management (n.d.) STREAM, cyber risk & compliance management platform. https://acuityrm.com/platform. Accessed 10 Feb 2021

46. CRR Supplemental Resource Guide Risk Management (n.d.). https://us-cert.cisa.gov/sites/def ault/files/c3vp/crr_resources_guides/CRR_Resource_Guide-RM.pdf. Accessed 10 Feb 2021

47. Common Controls Hub (n.d.) Compliance mapping for PCI, HIPAA, and more. https://com moncontrolshub.com. Accessed 10 Feb 2021

48. NIST (2020) Security and privacy controls for information systems and organizations. https:// nvlpubs.nist.gov/nistpubs/SpecialPublications/NIST.SP.800-53r5.pdf

49. public.cyber.mil (n.d.) Security technical implementation guides (STIGs)—DoD cyber exchange. https://public.cyber.mil/stigs. Accessed 10 Feb 2021

50. public.cyber.mil (n.d.) Community gold standard (CGS)—DoD cyber exchange. https://pub lic.cyber.mil/cgs. Accessed 10 Feb 2021

51. Cloud Security Alliance (n.d.) Cloud security alliance. https://cloudsecurityalliance.org/res earch/cloud-controls-matrix. Accessed 10 Feb 2021

52. ISO—International Organization for Standardization (2019) ISO/IEC 27001:2013. [online] ISO. https://www.iso.org/standard/54534.html.

53. 14:00–17:00 (n.d.) ISO/IEC 27701:2019. https://www.iso.org/standard/71670.html. Accessed 10 Feb 2021

54. 14:00–17:00 (n.d.) ISO/IEC CD 27402. [online] ISO. https://www.iso.org/standard/80136. html. Accessed 10 Feb 2021

55. Pcisecuritystandards.org (2019) Official PCI security standards council site—Verify PCI compliance, download data security and credit card security standards. https://www.pcisec uritystandards.org

56. CIS (2018) The 20 CIS controls & resources. https://www.cisecurity.org/controls/cis-controls-list
57. Isaca (2019) COBIT | control objectives for information technologies | ISACA. [online] Isaca.org. https://www.isaca.org/resources/cobit
58. BASELINE CYBER SECURITY CONTROLS FOR SMALL AND MEDIUM ORGANIZATIONS FOR SMALL AND MEDIUM ORGANIZATIONS. (n.d.) https://cyber.gc.ca/sites/default/files/publications/Baseline%20Cyber%20Security%20Controls%20for%20Small%20and%20Medium%20Organizations.pdf. Accessed 10 Feb 2021
59. www.ncsc.gov.uk (n.d.) About cyber essentials. https://www.ncsc.gov.uk/cyberessentials/overview
60. us-cert.cisa.gov (n.d.) Assessments: cyber resilience review (CRR) | CISA. https://us-cert.cisa.gov/resources/assessments. Accessed 10 Feb 2021
61. owasp.org (n.d.) OWASP application security verification standard. https://owasp.org/www-project-application-security-verification-standard
62. owasp.org (n.d.) OWASP mobile security testing guide. https://owasp.org/www-project-mobile-security-testing-guide
63. Zortrex (n.d.) Data protection—Secure tokenisation solutions. [online] Zortrex. https://www.zortrex.com. Accessed 10 Feb 2021
64. www.gcicom.net (n.d.) Gartner recognised contact centre solutions from GCI. https://www.gcicom.net/Our-Services/Unified-Communications/GCI-Contact-Centre. Accessed 10 Feb 2021

Implications of AI in National Security: Understanding the Security Issues and Ethical Challenges

Shasha Yu and Fiona Carroll

Abstract National security is the security and defence of a country. In the UK, national security provides coordination on security and intelligence issues of strategic importance across a government [25]. In the midst of the fourth industrial revolution, many countries including the UK have adopted Artificial Intelligence (AI) to achieve optimal national security for its citizens, economy, and institutions. In fact, AI, with its evolving intelligent behaviours, has become a major priority for defense for many governments. AI has the potential to support a number of domestic and international security initiatives, from online security to counter-terrorism. It also has the potential to change everything about the way government offices relate to each other and the rest of the world, how they work and ultimately get on together. Therefore, the more pressing question centres around how AI will shape national security? AI encompasses a series of complex issues that cut across security and ethical boundaries. This book chapter will discuss these themes in relation to national security. In detail, it will examine the security issues and ethical challenges of AI for the operational and strategic levels of national Security.

Keywords National security · Artificial intelligence · Security · Ethics · Technical

1 Introduction

From a societal perspective, Artificial Intelligence (AI) is already having huge impacts on our home life, workplaces, educational experiences, healthcare institutes, transport systems and military defense. As ([43], p.1) defines: 'AI is the ability of a computer or a robot to perform tasks commonly associated with intelligent beings'.

S. Yu · F. Carroll (✉)
Cardiff School of Technologies, Cardiff Metropolitan University Llandaff Campus, Western Avenue CF5 2YB, Cardiff, UK
e-mail: fcarroll@Cardiffmet.ac.uk
URL: https://www.cardiffmet.ac.uk/technologies/staff-profiles/Pages/Fiona-Carroll.aspx

S. Yu
e-mail: s.YU3@outlook.cardiffmet.ac.uk

© The Author(s), under exclusive license to Springer Nature Switzerland AG 2021   157
R. Montasari and H. Jahankhani (eds.), *Artificial Intelligence in Cyber Security: Impact and Implications*, Advanced Sciences and Technologies for Security Applications,
https://doi.org/10.1007/978-3-030-88040-8_6

The phenomenon of AI is truly in that a machine can think and act like a human (it can learn, problem solve, and reason like a human). It is believed that in conditions of 'unimaginable accumulation of information and the need for rapid decision-making, only the use of AI can lead to success' ([52], p.1). However, what does that mean for our national security? How can AI be implemented to handle some of the main challenges of national and global security? Challenges such as: 'Hybrid threats, economic crises, social inequalities, and labor migration' ([39], p.1528). Actually, in terms of national security, AI research is already underway in the fields of 'intelligence collection and analysis, logistics, cyber operations, information operations, command and control, and in a variety of semi autonomous and autonomous vehicles' ([56], p.10). While AI holds promise for addressing these challenges, it has to be noted that the rapid progress of AI and machine learning has opened up many new security issues and ethical challenges, particularly for the operational and strategic levels of national Security.

This chapter will take a deep look at these issues and challenges. The first section will give a snapshot of the national security picture for 2021. Following that, the chapter will focus on emerging AI related security issues such as digital, physical, political, economic and social security. It will then examine ethical challenges such as privacy concerns, machine bias, decision making, electronic personhood and legal responsibility. The chapter concludes with a reflection on the main points of interest from the research and then an insightful discussion looking to what the post pandemic future brings for AI in national security. In particular, how AI must always remain controlled as opposed to it taking control of our precious national security.

2   National Security in 2021

In 2021, the international and domestic security landscape has become increasingly complex and volatile. The ongoing world wide pandemic has hit the global economy hard, while cyber threats are becoming increasingly prevalent. These threats include both digital security threats caused by terrorists exploiting cyber vulnerabilities to launch hacking attacks, social security threats from people's online privacy leaks due to the widespread usage of social media, and political threats from the manipulation of social opinion by social bots. In addition to this, there is the economic threat of exploiting vulnerabilities to hack into national infrastructure networks, as well as the physical threat of using automated machines to conduct physical attacks.

A study in 2012 by EMC Digital Universe [26] estimated that global data would double every two years. In 2020, people created 1.7 MB of data every second [8]. Much of this new data is unstructured sensor or text data and stored in non-integrated databases, and such a huge amount of raw data is far beyond the limits of manual analysis to exploit. The exponential growth of digital data poses a huge technical challenge to national security, requiring the use of more sophisticated analytical tools to effectively manage risk in order to more proactively address emerging security threats. There is a growing need for AI to draw insights from all this data and to enhance the intelligence analysis work currently done by humans. AI has the capacity to find the finer details and connections in this data that could often go unnoticed by the human. However, AI also brings inherent risks. In recent times, there has been a lot of discussion on what threats artificial intelligence poses to human society. Bhatnagar et al. [7] classified the AI threats into three domains, which are digital security, physical security, and political security. Other scholars propose that the security threat needs to include: infrastructure security, endpoint security, web security, security operations and incident response, threat Intelligence, mobile security and human security [68]. In this book chapter, the authors focus on the digital, physical, political, economic and social security issues for national security endued by AI technologies.

2.1 AI and National Security

AI is a series of advanced general-purpose digital technologies that enable machines to effectively perform highly complex tasks [64]. Indeed, it is considered a strategic high point in the development of science and technology since the third industrial revolution. Moreover, it is believed that it will become a core technology for future economic and social development that will bring a profound changes to social, economic and industrial structures. AI is forecast to play a pivotal role in improving overall national competitiveness, and therefore, it is highly valued by all countries. This alone has hugely accelerated the pace of AI development. In the last decade, published AI patent applications have increased by 400% (Property Office 2019). The economic impact of robotics and autonomous cars will reach €650–120 billion per year by 2025, by which time AI and robotics could create 60 million new jobs worldwide [42]. In their PwC report [71], predict that the global market size of AI industry will reach $15.7 trillion by 2030.

In line with this, it is estimated that in 2016, North America invested €1218.6 billion in AI, Asia invested €6.5–€9.7 billion and Europe invested €2.4–€3.2 billion (far less than the former two) [4]. With the UK now officially leaving the EU, AI building is high on the agenda. The UK government announced in March 2021 that it is developing a new plan to become a global hub for responsible AI development, commercialization, and adoption. The National AI Strategy [25] is to be published in 2021. A study conducted by the Royal United Services Institute for Defence and Security Studies (RUSI) [5] on the use of AI for national security purposes shows that AI is being deployed primarily in the automation of administrative organizational processes, cybersecurity and intelligence analysis. Moreover, Allen and Chan [1] highlighted that advances in AI will impact national security by driving changes in three areas: military advantage, information advantage, and economic advantage. AI holds many opportunities to enhance national security however, with these do come many security concerns.

3   Security Issues

The rational for the use of AI is that it can promote social progress and liberate humans from monotonous and repetitive labour to engage in more creative work, thus contributing to the well-being of people. However, AI is also a double-edged sword, and unregulated misuse and abuse can easily lead to adverse consequences, and even catastrophic consequences. Therefore, the authors of this book chapter feel that it needs to be the combined responsibility of governments, research institutions, developers and users to ensure that the development and application of AI is on the right track. All these stakeholders need to work together to ensure sufficient transparency, strong oversight and accountability are ensured, both for the healthy development of technology and for national security. To achieve this, it is important to fully understand why AI poses such security risks?

The 'intelligence' of AI is such that it can automatically learn from known data to explore certain patterns and rules and apply them to new data. For the user, the corresponding inputs and outputs are clearly known, but the process is not understood, which is the 'black box effect' of AI processing [27]. The black box effect makes the application of AI uncertain and may lead to uncontrollable results. For example, on March 23, 2016, Microsoft Corporation released an artificial intelligence chatbot, called Tay, on Twitter. However, it was shut down after just sixteen hours because Tay quickly learned the inflammatory and offensive language that pervaded the internet, raising concerns about artificial intelligence [57].

The misuse of artificial intelligence also poses security threats. The increasingly widespread use of facial recognition technology in recent years has led to exposure of personal privacy [44]. What's more, some criminals maliciously use AI software to engage in criminal activities, such as imitating the voices of victims' friends and relatives to commit fraudulent activities by using voice synthesis software. In addition, accidents, such as power failures, data loss, or communication disruptions can also cause AI to go haywire. For example, drones crashing due to host failure etc. A website called 'Dronewars' [21] has done statistics on drone crashing from 2007 to 2020, with up to 400 publicly reported drone crashing from all causes, mostly due to mechanical or electrical failures. These drones, ranging in weight from micro and small drones under 150 kg to large drones over 600 kg, crash from high altitudes, posing a serious safety hazard.

Finally, biased or contaminated data are common causes of machine bias. Artificial intelligence learns from the data fed to it, so the quality of the data determines the effectiveness it learnt. Due to biased or contaminated data, AI may produce biased outputs. For example, when AI recommends jobs to job seekers based on their information, it may recommend lower-paying jobs to women based on the male–female salary gap in the database, even if the job seekers have excellent educational and professional backgrounds. This potential for bias can feed into a number of security concerns.

3.1 Digital Security

Nowadays, digital security and AI go hand in hand. Cognitive technologies (such as computer vision, machine learning, natural language processing, speech recognition, and robotics) are products of the field of artificial intelligence and are being found more frequently in security settings. For example, user access authentication, network situational awareness, dangerous behavior monitoring, abnormal traffic identification, etc. to protect network security [74]. Cognitive computing refers to systems that 'learn at scale, reason with purpose, and naturally interact with humans' ([37], p1.). This form of computing uses knowledge from cognitive science to build systems that mimic human thought processes and can process, understand, and add structured and unstructured information from a variety of sources to its knowledge base. Then, the cognitive system uses information based on this knowledge base to respond to complex problems. Applying cognitive technology to the cyber security domain can identify and respond to threats more quickly. For example, Watson for Cyber Security, IBM's AI cyber security system, uses security information from threat intelligence summaries, security events and related data [33]. It also makes use of data from unstructured sources such as research reports, security blogs, web sites and advisory reports. Currently, it has ingested more than 10 billion elemental components in its knowledge base and refreshes its understanding at a rate of more than 4 million elements per hour [17]. It also reads web data 24/7 and makes assumptions about attacks based on this dynamic knowledge base, reducing the time required for unexpected event analysis from hours to minutes, and it gets smarter as time goes on [17].

Establishing proactive security models helps to create the opportunities to disrupt and combat cybersecurity threats. The 'cyber kill chain' is a cybersecurity attack and defense model proposed by U.S. defense contractor Lockheed Martin [32]. This covers the seven phases required for a successful cyber attack: reconnaissance, weaponization, delivery, exploitation, installation, command control, actions on objectives. According to the model, the earlier the attack is stopped on the kill chain link, the lower the cost and time loss of remediation. Therefore, in the cyber security field, the earlier the intervention can bring better protection. Artificial intelligence (through machine learning and prediction) can provide early warning and can take blocking measures before a cyber attack occurs, establishing a proactive security model instead of the traditional reactive response approach. For example, AI^2 which is an AI-based cybersecurity platform developed by MIT, uses machine learning to autonomously scan data and activity, analyze network attacks, and then feed the findings back to cybersecurity analysts [35]. The cybersecurity analysts label the real cyberattack activities, feed the results back to the AI^2 system and apply them to new analyses. With this supervised learning, AI can rapidly improve prediction rates and increase accuracy as it learns more data, and it is ten times better at detecting attacks and five times less at alerting than other platforms. AI^2 is trained with over 360 million lines of log files, allowing it to analyze 85% of attacks in order to alert on suspicious behavior [15].

With these, there also comes the increase in the cases of AI misuse causing confidentiality, integrity and availability issues. Moreover, AI-powered malware is posing a growing threat to national security and people's privacy. For example, some hacker groups use AI to search for vulnerabilities in the nation's underlying information systems to automate cyber attacks. More criminal groups are being found to use voice synthesis software to impersonate victims' friends and relatives to commit fraud against them, or fake trusted email addresses to send advanced cyber pot phishing emails. Some even exploit the vulnerability of artificial intelligence systems to carry out adversarial attacks or disrupt the normal operation of artificial systems with data poisoning to cause errors, crashes, or even harmful consequences.

3.2 Physical Security

The wide and growing use of Internet of Things (IoT) technologies, autonomous cars, smart cities and interconnected national critical infrastructures will expose a large number of security vulnerabilities. These can be easily exploited to carry out attacks against humans and/ or the physical infrastructure impacting national wellbeing. For example, hackers can implement adversarial machine learning, tamper with data to trick neural networks, they can fool the systems into seeing what is not there and ignore what is there and even mis-classify objects. This can pose severe risks to public security.

Indeed, AI can have a huge impact on the physical world, for example, the field of autonomous cars. As ([30], p1) noted 'after adding perturbation to an original image, the AI could be manipulated to identify a 'stop' sign as a 'speed limit' sign'. In the field of sound recognition, adding specially designed noise disturbances to normal audio can allow a machine to recognize voice commands that are inaudible to humans, and execute a command as required [60]. Moreover, researchers did this experiment on the Samsung Galaxy S4 and iPhone 6, and successfully switched the phone into flight mode, dialed 911 and did other behaviors [11]. Malicious exploitation of this technology could lead to catastrophic physical security consequences, such as manipulating an autonomous car into committing a terrorist attack, or masquerade as a trusted user to bypass bio-metric authentication systems and perform hacking attacks.

3.3 Political Security

Artificial intelligence is not only powerful in data analysis, but also powerful in data production. Based on existing photos, videos, and audios, even amateurs can create high quality forged and synthetic media content, both for entertainment spoofing purposes and for political manipulation. The abundance of fake information affects the objective judgment of the public and even shakes the foundation of trust in the government and journalism. For example, on April 23, 2013, the official Twitter account of the Associated Press was hacked and fake news was released that 'Obama was injured in an explosion at the White House', which at once triggered panic in the market and the stock market plunged [63].

Furthermore, some interest groups use AI to create social opinion or carry out political attacks against their opponents. In the 2016 U.S. presidential election, 'social bots' began to emerge as an important factor in influencing U.S. politics on social media [36]. 'Social bots' are computer programs that automatically control social media accounts, scanning posts and comments of interest, and posting their own content to attract other human users. These botnets work in tandem to promote candidates and influence the public's political views and decisions [70]. Similar to social bots, deepfakes are also used for political purposes. Usually by replacing the face of a character in a video with the face of a politician or celebrity, there is the capability to decept (i.e. get them to say something they have never said). Deepfake uses Generative Adversarial Networks (GANs) technology to improve the algorithm's accuracy in replicating and fitting the real picture through continuous training [49]. This technique is like a game of cat and mouse. A network called the 'Generator' generates fake content based on real image data, and another network called the 'Discriminator' does the job of distinguishing real content from fake content. One produces 'lies' and the other discriminates 'lies', and this iterative process continues until the creator succeeds in tricking the discriminator.

Indeed, many researchers have now entered the fight against malicious deepfake, Microsoft has partnered with the Reality Defender tool and Google's Jigsaw has created its own tool called Assembler to reduce the harm spawned by malicous synthetic media [9, 29]. In addition, one of the solutions proposed by a team of researchers from Stanford University's Deepfake Research Team (DRT) is to use the same deep learning techniques as deepfake for inspection tools [69]. For example, Deepfake automated inspection tool Sherlock AI, which is an integrated convolutional model that looks for anomalous data information in videos. Developers say this tool can achieve 97% inspection accuracy [19]. Another approach to combat falsified media content is to use technologies such as blockchain to provide a safe and secure historical data tracking tool for the recognition of raw data. Blockchain is a completely new way to record Internet data, providing stable, tamper-evident data information in a decentralized, deconcentrated, distributed bookkeeping [59]. Vidprov, a blockchain program developed by the Stanford DRT group, attempts to provide proof of authenticity for video content by tracking the provenance of the video content [69]. Vidprov can track multiple edited versions of a video, and each video clip is tested to see if there is a smart contract associated with the parent video. If the parent video cannot be traced, then the digital video content cannot be trusted on the surface. With Vidprov, users can help determine whether a video can be traced to a trusted, reputable source, thereby combating deepfakes.

If we look to cyber terrorism, some extremist terrorist groups, such as the ISIS, use social media to recruit potential supporters of terrorist acts and expand their influence. They form dedicated teams to record propaganda videos, post tweets, images, videos, and set thematic hashtags to attract followers. Between 2010 and 2018, the UK's

Counter-Terrorism Internet Referral Unit (CTIRU) alone identified 300,000 pieces of terrorist content [40]. For this content proliferating on the web, Artificial intelligence can use image matching technology to control the upload of propaganda images or videos that were previously tagged as terrorist contents. The system can match a user's uploaded photo or video with a database of known terrorist information database to determine whether an upload is rejected. For example, YouTube's'redirect method' has been implemented to intelligently target and eliminate extremist propaganda from social media networks. The 'redirect method' detects ISIS-related searches, materials, ads and'related content' when they are detected, even'hidden' counter-argument content can be identified [53].

3.4 Economic Security

The widespread operation of artificial intelligence in the economy also poses pitfalls. Due to the high speed of autonomous systems, any small unexpected interaction and error can get out of control. One example is the May 2010 Wall Street stock market 'flash crash' in which a trillion dollars of stock market value was wiped out in a matter of minutes due to unexpected machine interactions (emergent effects) [62]. The American Securities and Exchange Commission (SEC) [55] reported that this event was enabled and exacerbated by the use of autonomous systems.

As far as the labour market is concerned, the widespread use of AI will inevitably cause a significant reduction in the overall demand for labour. As a result, a large number of unskilled workers will lose their jobs. Factory workers, construction workers, drivers, couriers, bank tellers, restaurant waiters, etc. will all be replaced by machines and/or software. Those countries and organizations that master advanced AI technologies will be able to make higher profits at lower costs. They will also be able to invest more capital in research and development. Thus initialising a virtuous circle in which the strong get stronger. The technology monopoly will make the social capital of certain national infrastructures stronger than others.

In this moment, technological strength will become an important indicator of national strength. In line with this, the importance of population size to national strength will be greatly reduced. Moreover, the large number of unemployed people who are replaced by AI and cannot update their knowledge and skills will become a heavy economic burden for the country.

3.5 Social Security

With the continuous development of artificial intelligence technology and the rapid improvement of computer computing power, the application of artificial intelligence will be more and more commonplace. Smart phones, tablets, personal computers, wearable devices and other kinds of intelligent terminals will allow users to connect to artificial intelligence application or platforms anytime and anywhere. As ([13], p.1) notes 'Artificial intelligence could contribute to the multi-pronged efforts to tackle some of the world's most challenging social problems'. However, with this potential for social good comes a variety of AI induced problems such as the production of fake photos, fake audio and fake video. As these proliferate through social media, it is difficult to distinguish what is true and what is false? These fakes could have a serious impact on social order and the building of social relations.

For para-journalism and self-published content creators, AI synthesized videos are inexpensive and mostly eye-catching, spreading quickly and widely. Whilst producing high-quality investigative journalism and verifying news sources is time-consuming and labour-intensive. As a result, the 'bad money drives out good money' effect can lead to fake news getting a head start and spreading faster and more widely. In some cases, even serious media may fail to distinguish between the real and the fake and in turn, also contribute to the spread of fake news [31]. Moreover, government's command and control organizations will face challenges. In more detail, those giving and receiving commands will have a difficult time knowing which communications (written, video. audio) are authentic. In 2019, the CEO of a British energy company was subjected to a phone scam [18]. The scammer used audio Deepfake technology to imitate the voice of the CEO of its parent company and instructed him to transfer €220,000 to a bank account in Hungary.

Moreover, the criminal justice field faces serious challenges in distinguishing between sources of information and authenticity of evidence, and even the most experienced experts can be fooled. In a study on the use of artificial intelligence to tamper with medical images, AI automatically injected or removed lung cancer information in a patient's 3D CT scan. The results were so convincing that they fooled three radiologists and the latest lung cancer detection AI [46]. The theft and misuse of photos and voices by others for fraudulent purposes will lead to a crisis of trust in human relationships. Organizations such as governments, banks, and commercial institutions will have to strengthen identity verification measures such as biometrics, which further increases the risk of personal information leakage, again affording a vicious cycle.

4 Ethical Challenges

The American writer Isaac Asimov proposed (in 1942) the famous 'Three Laws of Robotics' in his short story 'Runaround' [3]. This provided an early vision of the ethical rules for how robots could live in harmony with humans. Since then, humans have never stopped exploring the ethics of AI, and in 2017, the Asilomar 23 Guidelines for Artificial Intelligence were developed [34]. Now, the principles have been signed by 1,797 AI/robotics experts and 3,923 researchers. However, humanity continues to face enormous challenges with regard to the outputs of AI.

4.1 Privacy Concerns

Machine learning-based biometrics has gained significant momentum in recent years. A large number of open and trained machine learning libraries have made technologies such as facial recognition readily available. Some governments and organizations are using this technology for personnel surveillance, exposing personal privacy to ubiquitous surveillance networks [16]. Businesses use intelligent recommendation systems based on user behavior analysis to push specific products and information to targeted customers. This can violate users' rights to free access to information and even narrowing the window of access to information due to the 'echo chamber' effect [14].

To address the impact of this, Stanford University scholars used deep neural networks to extract features from 35,326 facial images for analysis to distinguish between homosexuals and heterosexuals, achieving an accuracy rate of 81% for men and 74% for women [72]. This has to raise concerns that in the face of artificial intelligence, even private matters like sexual orientation is clearly written on the face, what else can be hidden from AI? What's more, in the study the authors also suggest that artificial intelligence could be used to explore the links between facial features and a range of other phenomena, such as political views, psychological conditions or personality [72]. It is worrisome to think that the personal photos posted by social media users on the platform could in the future become a window for others to pry into their privacy.

4.2 Machine Bias

The AI learning process is the result of a combination of humans, data, and algorithms, and problems in any one of them may lead to bias. We classify biases into three categories based on what causes them: 'human-induced bias', 'data-driven bias' and 'machine self-learning bias'. Firstly, AI engineers may be unconsciously influenced by personal values, beliefs, and scope of knowledge when developing AI products, which may cause bias. When an engineer provides an AI model with a training set that is missing a feature or features for the above reasons, the AI-acquired model is unable to measure this part of the features correctly. In addition, in models with supervised learning, the AI needs to learn from the provided labels. In this process, human bias is inevitably brought into it unintentionally. For example, if a person labels a tall, slim person as 'beautiful', the algorithm will naturally inherit this bias.

Secondly, the learning principle of artificial intelligence is based on the analysis of existing data. To a large extent, the output results of artificial intelligence are determined by its training data, so data is the core asset of artificial intelligence. However, some of the data from real life are inherently imprinted with human bias, thus leading to the creation of machine bias. For example, predictive policing forecasts the likelihood of cases by analyzing data such as the number of cases in the community and the characteristics of offenders. This allows police to strengthen their presence in key areas and improve the efficiency of police resource utilization [45]. However, such analysis can label households in specific areas and specific human races as 'high-risk', undermining the principle of presumption of innocence.

Thirdly, Artificial intelligence learning is a self-exploratory, self-correcting, and ever-changing process in which bias can result from incorrectly inferring causality or correlation from something that is contingently connected. For example, when researchers trained a machine learning model to identify wolves and dogs, the AI automatically identified 'dog' as being more closely related to 'grass' and 'wolf' as being more closely related to 'snow' and identified the dog in the snow as a wolf based on this [6].

4.3 Decision Making

People are emotional creatures, while machines can make quick decisions through completely rational analysis. But when humans leave decision making to machines for the sake of efficiency, they often sacrifice fairness and justice, which is what humanity holds most precious. When humans face moral dilemmas, such as the Trolley Problem, different people will make different judgments based on their own value beliefs [61]. The Massachusetts Institute of Technology (MIT) built a website called the 'Moral Machine' [41] which uses the background of a driverless car with sudden brake failure that must kill different people in an emergency on the road to let users make a choice. The result is that people of different countries, nationalities and beliefs will potentially make completely different choices. However, when we are confronted with this problem in a fully automated vehicle in real life, it is neither the owner nor the passengers who makes the decision, but rather the algorithm of the AI or the data used. In other words, we are dependent on the developer of the AI. In this case, human lives are manipulated, and values set by machines override human values.

In July 2020, students in the UK were not able to take A-level and GCSE exams due to the epidemic [10]. Instead a computer program was used to assess student performance. This was based on students' performance in previous exams, as well as the progression record of the school they attended over the past few years. As a result, the model ended up favoring students in private schools and affluent districts, while free, high-achieving students in public schools were disproportionately affected. The automated algorithm lowered the A-Level scores of nearly forty percent of students, with high-achieving students from minority and low-income neighborhoods being the most affected. Protests over admissions discrimination eventually led to a UK government U-turn on grading methods.

4.4   Electronic Personhood

As artificial intelligence becomes more and more intelligent, whether we should recognize the electronic personality of AI becomes a question we must face. In October 2017, Sophia, a humanoid robot developed by Hanson Robotics (Hong Kong) [28], became a citizen of Saudi Arabia, the first robot in the world to receive citizenship. She can learn and adapt to human behavior, work with humans, and be interviewed around the world. In November 2017, Sophia was named the first non-human innovation champion by the United Nations Development Programme (UNDP) [48]. With this, the question of the robot's personality has sparked intense discussion [24, 54].

This discussion has focused on responsibility, conscience, and motivation. Some questions include: Should robots have personhood and legal identity like humans? To what extent can robots have the same status? Can self-learning robots be held accountable for their actions? Should they be held responsible for harming people or destroying property? How should the IP of inventions developed by AI agents be protected? Is the attribution of 'electronic personalities' the answer? These are the legal and ethical questions before us.

Some argue that the existing legal system cannot solve the emerging legal problems (across various fields) brought about by the increasingly popular robotic applications. Moreover, some see the granting of a legal personality to robots to be the answer (i.e. it can make them a clear legal subject). Indeed, supplemented by insurance and other measures, they feel that it can solve the legal problems involving AI such as legal recourse and compensation under the existing legal framework. They believe that robots can assume the same legal responsibilities and obligations as other legal persons such as companies, organizations, and associations.

Opponents, however, argue that creating a legal personality for robots would reduce the responsibility of the robot maker. To the extent, that the responsibility due to the maker is shared by the robot. They feel that this could enable manufacturers to circumvent the liability that they create. Furthermore, they also assert that robots have no human conscience and that their only task is to serve humans. Recognizing robotic electronic personhood raises many difficult ethical questions, such as whether robots can marry or vote like humans, whether it is considered murder for humans to intentionally shut down robotic systems, and so on. Another issue is that if robots are allowed to have personality, the works they (robots) create would have the same rights as human works, which would be detrimental to motivating human creators.

At this stage, creating specific legal rights and responsibilities for AI agents, rather than human rights, seems to be a more feasible approach. In October 2020, the European Parliament issued three resolutions on the ethical issues, civil liability and intellectual property rights of artificial intelligence software systems ('AI') [4]. All three resolutions insist on not providing legal personality to AI software systems.

4.5 Legal Responsibility

Along with the issue of robot personality comes the issue of legal liability [51]. The application of artificial intelligence involves developers, managers, users, algorithms, data etc. and robots' mistakes and/or crimes are often the result of intertwined stakeholders and causes. Who, then, should be held responsible for the faults and failures of robots? As a result of the 'black box effect' of AI, people cannot fully grasp the process of AI; they cannot understand who or what is responsible for the benefits and harms of using artificial intelligence [47]. In terms of the harms, when AI causes accidents or disasters, it is often difficult to determine the cause of the damage. Moreover, due to the autonomy of AI learning, two executions of the same thing may produce different results, so it is often difficult to accurately recover the process of the harm. This poses a huge challenge to both the investigation and the collection of evidence.

Moreover, large-scale AI systems are often the result of multiple parties working together and in close collaboration. In this case there is great uncertainty in the allocation of responsibility. Legal recourse to crime usually takes into account the subjective factors. For example, whether the perpetrator was intentional or negligent etc. However, in the case of AI, judging the subjective factors becomes more difficult. It is an ambiguous question whether the use of AI is malicious or the risks are purely associated with bold innovation. On the one hand, the development of human technology needs to motivate and enable scientists to keep trying to explore. On the other hand, to prevent criminal acts from being committed by the hands of unsuspecting people through AI.

A more rational regulatory system would ensure that AI remains on a healthy trajectory. The use of AI is characterized by opacity, complexity, reliance on data, and autonomous behavior, and the regulation of it is therefore complex. In April 2021, the EU published a proposal for a Regulation of the European Parliament and of the Council establishing uniform rules on artificial intelligence (AI Act) and amending certain Union legislation to regulate the use of artificial intelligence [22]. The proposal sets out how companies and governments should use AI technologies. The proposal implements different regulatory measures based on different risk levels and is considered a better option than comprehensive regulation of all AI systems. This framework distinguishes the use of AI as unacceptable risk, high risk, low risk, or minimal risk. The types of risks and threats are based on a sector-by-sector and case-by-case approach.

5 Discussion

The COVID-19 pandemic has presented many challenges and also opportunities for the accelerated development and application of AI. For example, hospitals and medical institutions worldwide are now using it for test-free diagnostic screening of COVID patients [58]. Pharmaceutical companies are developing AI to advance vaccines and new drugs [38]. Social isolation and lockdown are driving more and more robots and AI software into use, taking over workers in housekeeping, transportation, logistics, communications, education, manufacturing, and many other industries where many people have lost their jobs. According to the UnitedNations [67] the global economy is experiencing it's worst recession in ninety years, with an estimated one hundred and fourteen million jobs lost globally, some one hundred and twenty million people trapped in extreme poverty and greater inequality between countries. The impact has been sudden yet profound. As a result, the authors of this chapter, feel that in the post-epidemic era, countries will most certainly turn to AI to accelerate the development and application of high technology in order to recover their economies.

In line with this, it is estimated that the global AI market is expected to continue to grow at a compound annual growth rate of 42.2% from 2021 to 2027 [73]. As computing power continues to increase and the use of big data, 5G, cloud and IoT technologies become more commonplace in society, artificial intelligence will also become more accessible to the general public. In fact, evolving technologies such as augmented intelligence, cognitive technology, explainable AI (XAI), codeless or low-code machine learning, and automatic machine learning (AutoML) will also play a role in heightening the intrigue of artificial intelligence. Indeed, codeless, self-discovering, self-training, and self-managing platforms such as Microsoft's Lobe will enable people with no knowledge of code to create their own AI applications and deploy them easily [40]. AI has all the ingredients to become a crucial factor in the make-up of our human society.

However to make this work, international organizations and industry groups are developing guidelines, norms and standards to guide the development of AI within an ethical framework. The United Nations Educational, Scientific and Cultural Organization (UNESCO) published the COMEST report on robotic ethics in 2017. They also launched the Global Recommendation on the Ethics of Artificial Intelligence in November 2019; this is expected to be finally adopted at the UNESCO General Conference in November 2021 [66]. The Recommendation proposes four values, namely respect; protection and promotion of human dignity; human rights and fundamental freedoms; Environment and ecosystem. The Recommendation calls for upholding principles of proportionality. It is an advocate for safety, security, fairness, non-violence, non-discrimination, sustainability, privacy, human oversight and determination, transparency and explainability, responsibility and accountability, awareness and literacy, multistakeholder and adaptive governance and collaboration [66]. The recommendation calls for the implementation of policy actions in ten areas: ethical impact assessment, ethical governance and stewardship, data policy, development and international cooperation, environment and ecosystems, gender, culture, education and research, economy and labour, health and social wellbeing [66].

Moreover, the Institute of Electrical and Electronics Engineers (IEEE) released two editions of 'Ethically Aligned Design', issuing initiative on the Ethics of Autonomous and Intelligent Systems(A/IS) [12]. This global initiative encourages any individual or group involved in the research, design, manufacture, or exchange of information about autonomous and intelligent systems, including the universities, institutions, governments, and businesses that implement these technologies, to prioritize ethical issues in the A/IS development process. It proposes that the ethical design, development, and application of AI technologies should be guided by the principles of human rights, well-being, accountability, transparency, and prudent use. In March 2019, the European Union Agency for Cybersecurity (ENISA) [24] published a report titled 'Towards a framework for policy development in cybersecurity—Security and privacy considerations in autonomous agents'. This aims to provide EU member states with a framework for security and privacy policy development provision to address the growing popularity of AI applications. In addition, the European Commission released in April 2019, the official version of Ethics guidelines for trustworthy AI, which proposes a framework for achieving a full lifecycle of trustworthy AI [20]. The book defines 'trustworthy AI' as consisting of two essential components: first, it should respect basic human rights, regulations, core principles and values; and second, it should be technologically safe and secure to avoid unintentional harm caused by inadequate technology.

In terms of national development, major countries have now entered a new strategic era of comprehensively advancing the development of artificial intelligence. Canada, Japan, Singapore, China, the UAE, Finland, Denmark, France, the UK, the EU, South Korea and India have all issued policies to promote AI applications and development. Indeed, national strategies have become centred around science and innovation. In October 2017, the UK government released a report on 'Growing the Artificial Intelligence Industry in the UK' [64]. This report analyzed the current application, market and policy support of artificial intelligence, and proposed promotion from four aspects: improve access to data, improve supply of skills, maximise AI research and commercialisation, support the uptake of AI [64]. The report has been an important guideline for the development of artificial intelligence in the UK. Furthermore, in April 2018, the UK government released the 'AI Sector Deal' [65], which aims to promote the United Kingdom to become a global AI leader. This includes promoting government and company research and development, STEM education investment, upgrading digital infrastructure, increasing AI talents, and leading global digital ethics exchanges etc. [65].

Emerging from these initiatives, organizations and industries are taking active measures to capitalise on the opportunities but also to deal with the challenges that artificial intelligence affords. Indeed, critical infrastructure systems such as transportation, power, banking, communications, and medical care realise that they need to proactively embrace AI technology to prevent disasters and improve safety and security. National culture and education departments are realising that they need to put the training of high-level artificial intelligence talents and the popularization of knowledge on a level playing field in order to nurture a sustainable reserve of experts. Human resources and labour departments are recognising the need to make forward-looking predictions for future industry changes, implement effective employment guidance and guarantees, and respond to the impact of labour transfer and unemployment caused by AI. Scientific research institutions are inspiring the development of ethical frameworks whilst also accelerating the pace of research. Furthermore, the legal sectors are working to ensure feasible legal frameworks so that the design, production, and use of artificial intelligence are within effective supervisory frameworks and legislature. Finally, the insurers are positioning themselves to explore new insurance models and incorporate artificial intelligence into network insurance plans to respond to new market needs. All in all, governments are realising that the development of artificial intelligence technology is a valuable asset to a countries prosperity as well as it's security. However, they are also understanding that there needs to be joint efforts from international organisations, countries, societies and individuals to ensure that AI is designed effectively and works for both national welfare and security.

6 Conclusion

Since John McCarthy first coined the term 'artificial intelligence' in 1956 (over 65 years ago), artificial intelligence has developed from a reasonably unknown concept to now a common day term which infiltrates all aspects of our work and life [2]. Just as machines, electricity and computers played a decisive role in the previous industrial revolution, in the era of Industry 4.0, a country's artificial intelligence development level determines a country's security and future development. As we discussed in this book chapter, artificial intelligence poses serious challenges in terms of digital security, physical security, political security, economic security, and social security. Its application in industry and in life also has ethical challenges. As we have seen, under the current technical levels and legal frameworks, there are also some difficulties in investigating and collecting evidence for errors or crimes caused by artificial intelligence.

However, we must realize that while artificial intelligence creates issues and brings many challenges, it is also accompanied by huge opportunities. Making full use of artificial intelligence technology and establishing a reasonable legal framework and regulatory system, it will certainly be able to advance to be a powerful tool to promote national security and economic development. As discussed the continuous development of appropriate systems (especially the ethical regulations and standards that countries and international organizations are establishing) have created a good foundation for us to face the challenges and mould AI to better serve a nation and create a brighter and more secure future for everyone.

References

1. Allen G, Chan T (2017) Artificial intelligence and national security. Tech. rep. www.belfercenter.org
2. Andresen SL (2002) John McCarthy, Father of AI
3. Asimov I (1942) Runaround. Astounding science fiction

4. Ayet Puigarnau J (2018) Communication from the commission to the European Parliament, the European council, the council, the European Economic and social committee and the committee of the Regions. https://data.consilium.europa.eu/doc/document/ST-8507–2018-INIT/en/pdf
5. Babuta A, Oswald M, Janjeva A (2020) Artificial intelligence and UK national security: policy considerations. Tech. rep
6. Besse P, Castets-Renard C, Garivier A, Loubes JM (2018) Can everyday AI be ethical. Fairness of Mach Learn Algorith. arxiv.org/abs/1810.01729
7. Bhatnagar S, Cotton T, Brundage M, Avin S, Clark J, Toner H, Eckersley P, Garfinkel B, Dafoe A, Scharre P, Zeitzoff T, Filar B, Anderson H, Roff H, Allen GC, Carrick JS, Sèan F, Hèigeartaigh O, Beard S, Belfield H, Farquhar S, Lyle C, Crootof R, Evans O, Page M, Bryson J, Yampolskiy R, Amodei D (2018) The Malicious use of artificial intelligence: forecasting, prevention, and mitigation authors are listed in order of contribution design direction, vol 101. https://img1.wsimg.com/blobby/go/3d82daa4–97fe-4096–9c6b-376b92c619de/downloads/1c6q2kc4v_50335.pdf. arXiv:1802.07228
8. Bulao J (2021) How much data is created every day in 2021? [You'll be shocked!]. https://techjury.net/blog/how-much-data-is-created-every-day/
9. Burt T, Horvitz E (2020) New steps to combat disinformation—Microsoft on the issues. https://blogs.microsoft.com/on-the-issues/2020/09/01/disinformation-deepfakes-newsguard-video-authenticator/
10. Busby M (2020) A-level student launches legal bid against of qual—A-levels—The guardian. https://www.theguardian.com/education/2020/aug/16/a-level-student-launches-legal-bid-against-ofqual
11. Carlini N, Mishra P, Vaidya T, Zhang Y, Sherr M, Shields C, Wagner D, Zhou W (2016) Hidden voice commands. In: Proceedings of the 25th USENIX security symposium, pp 513–530
12. Chatila R, Havens JC (2019) Ethically aligned design (IEEE)
13. Chui M, Harrysson M, James Manyika J, Roberts R, Chung R, Nel P, van Heteren A (2018) Notes from the AI frontier. Applying AI for social good. McKinsey Global Institute
14. Cinelli M, de Francisci Morales G, Galeazzi A, Quattrociocchi W, Starnini M (2021) The echo chamber effect on social media. In: Proceedings of the National Academy of Sciences of the United States of America
15. Conner-Simons A (2016) System predicts 85% of cyber-attacks using input from human experts—MIT News—Massachusetts Institute of Technology. https://news.mit.edu/2016/ai-system-predicts-85-percent-cyber-attacks-using-input-human-experts-0418
16. Connor BT, Doan L (2021) Government and corporate surveillance: moral discourse on privacy in the civil sphere. Inf Commun Soc
17. Crume J, Lhotka D, Austin C (2018) Security and artificial intelligence. Tech. rep. https://en.wikipedia.org/wiki/Artificial_intelligence
18. Damiani J (2019) A voice deepfake was used to scam a CEO out of $243,000 (2019). https://www.forbes.com/sites/jessedamiani/2019/09/03/a-voice-deepfake-was-used-to-scam-a-ceo-out-of-243000/?sh=192aef622241
19. DeepQuanty: Sherlock AI—DeepQuanty (2020). http://deepquantyailabs.com/sherlock-ai/
20. Directorate-General for Communication: the EU in 2019—General Report on the Activities of the European Union (2020). https://op.europa.eu/webpub/com/general-report-2019/en/#chapter0
21. Dronewars: Drone Wars UK (2021). https://dronewars.net/
22. EU: EUR-Lex-32021R0240-EN-EUR-Lex (2021). https://eur-lex.europa.eu/eli/reg/2021/240
23. European Union Agency for Cybersecurity: Mehari—ENISA (2020)
24. Fosch-Villaronga E Lutz C, Tamò-Larrieux A (2020) Gathering expert opinions for social Robots' Ethical, legal, and societal concerns: findings from four international workshops. Int J Soc Robot
25. GOV.UK: National security and intelligence—GOV.UK (2021), https://www.gov.uk/government/organisations/national-security.
26. Gantz J, Reinsel D (2012) IDC the digital Universe in 2020: big data, bigger digital shadows, and biggest growth in the far east sponsored by EMC Corporation. Tech. rep. www.emc.com/leadership/digital-universe/index.htm

27. Guidotti R, Monreale A, Pedreschi D (2019) The AI black box explanation problem. Ercim News
28. Hanson-Robotics: Sophia-Hanson Robotics (2020). https://www.hansonrobotics.com/sophia/
29. Hao K (2021) Google has released a tool to spot faked and doctored images. MIT Technol Rev. https://www.technologyreview.com/2020/02/05/349126/google-ai-deepfakes-manipulated-images-jigsaw-assembler/
30. Heaven D (2019) Why deep-learning AIs are so easy to fool. Nature
31. Hofseth, A.: Fake news, propaganda and influence operations – a guide to journalism in a new and more chaotic environment (2017)
32. Hutchins E (2018) Cyber Kill Chain®—Lockheed Martin
33. IBM: Artificial Intelligence for Smarter Cybersecurity—IBM (2021). https://www.ibm.com/security/artificial-intelligence
34. Institute FoL (2020) Asilomar AI principles. Future of Life Institute, Understanding the Security issues and Ethical challenges 19
35. Iyer KS (2016) MIT researchers develop AI that can detect cyberattacks with 85% accuracy TechWorm. https://www.techworm.net/2016/04/mit-researchers-develop-ai-detect-cyberattacks-85-accuracy.html
36. Jacobs S (2018) Political Propaganda spread through social bots. In: Summer lightfoot
37. Kelly JE (2015) Computing, cognition and the future of knowing. IBM White Paper
38. Keshavarzi Arshadi A, Webb J, Salem M, Cruz E, Calad-Thomson S, Ghadirian N, Collins J, Diez-Cecilia E, Kelly B, Goodarzi H, Yuan JS (2020) Artificial intelligence for COVID-19 drug discovery and vaccine development. Front Artif Intell
39. Limba T, Stankevičius A, Andrulevičius A (2019) Industry 4.0 and national security: the phenomenon of disruptive technology. Entrep Sustain Issues
40. Luyten K (2021) Lobe: Lobe—Machine Learning Made Easy, Addressing the dissemination of terrorist content online. Tech. rep. https://www.lobe.ai/
41. MIT: Moral Machine (2016). https://www.moralmachine.net/
42. Manyika J, Chui M, Bughin J (2013) Disruptive technologies: advances that will transform life, business, and the global economy. McKinsey Global
43. Martin AS, Freeland S (2021) The advent of artificial intelligence in space activities: new legal challenges. Space Policy
44. Martin N (2019) The major concerns around facial recognition technology. https://www.forbes.com/sites/nicolemartin1/2019/09/25/the-major-concerns-around-facial-recognition-technology/?sh=29b3bafb4fe3
45. Meijer A, Wessels M (2019) Predictive policing: review of benefits and drawbacks. Int J Public Adm
46. Mirsky Y, Mahler T, Shelef I, Elovici Y (2019) CT-GAN: malicious tampering of 3D medical imagery using deep learning
47. Neri E, Coppola F, Miele V, Bibbolino C, Grassi R (2020) Artificial intelligence: who is responsible for the diagnosis?
48. Pagallo U (2018) Vital, Sophia, and Co.-The quest for the legal personhood of robots. Information (Switzerland)
49. Pan Z, Yu W, Yi X, Khan A, Yuan F, Zheng Y (2019) Recent progress on generative adversarial networks (GANs): a survey. IEEE Access
50. Property Office I (2019) Artificial intelligence
51. Rachum-Twaig O (2020) Whose robot is it anyway?: liability for artificial intelligence-based robots. University of Illinois Law Rev
52. Radulov N (2019) Artificial intelligence and security. Instr Sci J "Security and Future"
53. Redirect: The Redirect Method (2021). https://redirectmethod.org/
54. Robert LP, Alahmad R, Esterwood C, Kim S, You S, Zhang Q (2020) A review of personality in human-robot interactions
55. SEC: SEC.gov—HOME (2019). https://www.sec.gov/
56. Sayler KM (2020) Artificial intelligence and national security—Economic impacts and considerations. Congr Res Serv

57. Schwartz O (2019) In 2016, microsoft's racist chatbot revealed the dangers of online conversation
58. Soltan AA, Kouchaki S, Zhu T, Kiyasseh D, Taylor T, Hussain ZB, Peto T, Brent AJ, Eyre DW, Clifton DA (2021) Rapid triage for COVID-19 using routine clinical data for patients attending hospital: development and prospective validation of an artificial intelligence screening test. Lancet Dig Health
59. Song G, Kim S, Hwang H, Lee K (2019) Blockchain-based Notarization for social media. In: 2019 IEEE international conference on consumer electronics, ICCE 2019 (2019)
60. Sugawara T, Cyr B, Rampazzi S, Genkin D, Fu K (2019) Light commands: laser-based audio injection attacks on voice-controllable systems
61. Thomson JJ (1985) The Trolley problem. Yale Law J
62. Treanor J (2015) The 2010 'flash crash': how it unfolded—Stock markets—The Guardian. https://www.theguardian.com/business/2015/apr/22/2010-flash-crash-new-york-stock-exchange-unfolded
63. Truther: Cyber Terror Attacks (2013) TRUTHER.ORG: exposing false-flag terror since 2007 (2021). https://sites.google.com/site/truthersswitzerland/cyber-terror/cyber-terror-attacks-2013
64. UK Government: Growing the artificial intelligence industry in the UK: recommendations of the review. Gov.uk (2017)
65. UK Government: AI Sector Deal—GOV.UK (2019). https://www.gov.uk/government/publications/artificial-intelligence-sector-deal/ai-sector-deal
66. UNESCO: outcome document: first draft of the Recommendation on the Ethics of Artificial Intelligence—UNESCO Digital Library (2020). https://unesdoc.unesco.org/ark:/48223/pf0000373434
67. UnitedNations: The global COVID-19 pandemic may lead to the loss of a decade of development gains—UN News (2021). https://news.un.org/zh/story/2021/03/1080892
68. Vähäkainu P, Lehto M (2019, February) Artificial intelligence in the cyber security environment. In: ICCWS 2019 14th International Conference on Cyber Warfare and Security: ICCWS 2019. Academic Conferences and Publishing Limited, p 431
69. Van de Weghe T (2019) Six lessons from my deepfakes research at Stanford—by Tom Van de Weghe—JSK class of 2019—Medium. https://medium.com/jsk-class-of-2019/six-lessons-from-my-deepfake-research-at-stanford-1666594a8e50
70. Vasilkova VV, Legostaeva NI (2019) Social bots in political communication. RUDN J Sociol
71. Verweij G, Rao A (2017) Sizing the prize: what's the real value of AI for your business and how can you capitalise? PwC
72. Wang Y, Kosinski M (2017) Deep neural networks can detect sexual orientation from faces. J Person Soc Psychol 1–47
73. Webb A (2021) Tech Trends Report—The future today institute. https://futuretodayinstitute.com/trends/. Understanding the Security issues and Ethical challenges 21
74. Zhang Z, Ning H, Shi F, Farha F, Xu Y, Xu J, Zhang F, Choo KKR (2021) Artificial intelligence in cyber security: research advances, challenges, and opportunities. Artif Intell Rev

DeepFakes: Detecting Forged and Synthetic Media Content Using Machine Learning

Sm Zobaed, Md. Fazle Rabby, Md. Istiaq Hossain, Ekram Hossain, Md. Sazib Hasan, Asif Karim, and Khan Md. Hasib

Abstract The rapid advancement in deep learning makes the differentiation of authentic and manipulated facial images and video clips unprecedentedly harder. The underlying technology of manipulating facial appearances through deep generative approaches, enunciated as *DeepFake* that have emerged recently by promoting a vast number of malicious face manipulation applications. Subsequently, the need of other sort of techniques that can assess the integrity of digital visual content is indisputable to reduce the impact of the creations of DeepFake. A large body of research that are performed on DeepFake creation and detection create a scope of pushing each other beyond the current status. This study presents challenges, research trends, and directions related to DeepFake creation and detection techniques by reviewing the notable research in the DeepFake domain to facilitate the development of more robust approaches that could deal with the more advance DeepFake in future.

S. Zobaed · Md. F. Rabby · E. Hossain
University of Louisiana, Lafayette, LA 70503, USA
e-mail: sm.zobaed1@louisiana.edu

Md. F. Rabby
e-mail: sourav.sust.cse.10@gmail.com

E. Hossain
e-mail: ekram.hossain1g@louisiana.edu

Md. I. Hossain
Southern Utah University, Cedar City, UT 84720, USA

Md. S. Hasan
Dixie State University, St. George, UT 84770, USA
e-mail: mdsazibhasan@dixie.edu

A. Karim (✉)
Charles Darwin University, Casuarina, NT 0810, Australia
e-mail: asif.karim@cdu.edu.au

K. Md. Hasib
Ahsanullah University of Science Technology, Dhaka, Bangladesh

© The Author(s), under exclusive license to Springer Nature Switzerland AG 2021   177
R. Montasari and H. Jahankhani (eds.), *Artificial Intelligence in Cyber Security: Impact and Implications*, Advanced Sciences and Technologies for Security Applications,
https://doi.org/10.1007/978-3-030-88040-8_7

Keywords DeepFake generation · DeepFake detection · Adversarial attack · Face swap

1 Introduction

Because of the advances of deep learning and generative adversarial networks (GAN) [1], creation of a realistically looking face image of a target person who really does not exist or altercation of facial appearance (attributes, identity, expression) is attainable with maintaining realism. The deep learning research community roughly refers to the technology as "DeepFake" that is coined from "deep learning" and "fake". Generally, DeepFake approaches require a massive volume of image and video data to train models for generating realistic images and videos. Because of the wide availability of robust pretrained DeepFake mod els, malicious DeepFake contents are generated that create negative impact on the societies.

The potential target of DeepFake is the public figures such as celebrities, priests, and politicians whose videos and images are largely available on the internet. More specifically, DeepFake is often used to alter faces of celebrities or politicians to other bodies in pornographic contents. DeepFakes can be abused to create political or ethnic tensions between countries to fool common people to affect election, or create chaos in sports or global economy by creating fake contents.

There are numerous notable examples of DeepFake incidents that have been shared in the internet [2–5]. For instance, in 2018, a video posted in Facebook showing Former President of USA, Donald Trump taunted Belgium for remaining in the Paris climate agreement [4]. By noticing the video clearly, it was determined that Trump's hair looked stranger than usual and his voice was rolled up. In 2019, a DeepFake video of Facebook owner, Mark Zuckerberg, was published on Instagram [4]. In the video, Zuckerberg's speech was altered along with his facial expression so that the viewers can easily be distracted. A recent release of an app named DeepNude raises issue since it is used to transform a person to a non-consensual pornography [6]. Similarly, a Chinese app named "Zao" got viral lately for offering face swapping with bodies of TV stars and even replace themselves into wellknown movies and TV clips [7]. These forms of manipulation create a serious threat to privacy and identity, and even jeopardize personal lives. Although the evil technology is undoubtedly a severe threat to world security, it is also used in positive purposes such as updating episodes of a visual content even after the actor is dead or creating speech of mute people. However, the number of maliciously used cases DeepFake significantly outperforms the number of positively used cases.

The underlying mechanism for DeepFake creation is advanced deep learning models such as autoencoders and GAN, which have been applied widely in the computer vision research community. Due to the development of advanced deep networks and the availability of a substantial amount of training data, manipulated images and videos have turned out almost indistinguishable to human eyes and even to robust algorithms. Hence, the creation of those manipulated contents becomes simpler and takes comparatively lesser effort. This is because an identity image or small video clip of a targeted individual are sufficient for the inference tasks.

The rise of stunning DeepFake creation vividly highlights the significance of judging the genuineness of digital media content. Because of the availability of various DeepFake creation tools, almost anyone can simply create forged content these days. As a result, in the computer vision research community, the study of DeepFake has gained traction in recent years for detecting such contents [8–12]. In [10], Juefei-Xu et al. showed a distribution of DeepFake related papers in last 5 years, where 78% of the total papers published in the last two years. This increase amount of paper in the last two years vividly highlights research interest revolved around DeepFakes.

Governments and law enforcement are undertaking the spread of DeepFake creations with new policies and regulations as well. For example, US Senator named. Ben Sasse proposed a bill titled *S.3805—Malicious deep fake prohibition act of 2018* in 2018 that introduces a new type of criminal offense because of the creation or distribution of fake digital media content that falsify realism [13]. Besides, social media platforms (e.g., Twitter, Facebook) are actively taking initiatives to deal with forged, synthetic, and manipulated content on their respective platforms. For example, in Twitter, if a tweet contains manipulated media content specially, DeepFakes, Twitter has started to alert users about that by tagging with warning sign and attaching the trustworthy news article link relevant to the tweet [14]. In another example, in 2019, Facebook facilitated the development of robust DeepFake detection tools by organizing the DeepFake detection challenge (DFDC) where 2114 number of participants across the globe had participated and they generated more than 35,000 models [15].

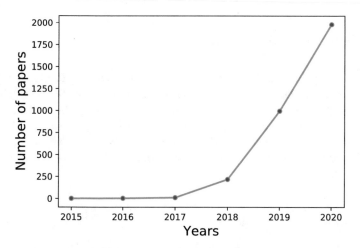

Fig. 1 Number of scholarly articles that are related to DeepFake in years from 2015 to 2020, obtained from Google Scholar on April 2021 with the query keyword "deepfake" applied to either title or full text of the papers

In Fig. 1, we depict the relation between number of papers that are re lated to DeepFake in years from 2015 to 2020. The data is collected from Google Scholar on April 2021 with the query keyword "deepfake" found in either title or full text of the papers. According to the number of related papers has increased significantly in the recent years which is an indication Deep Fakes related research or news are getting noticed a lot more in recent times.

This chapter presents a handful number of methods for DeepFake generation and detection in comprehensive manner. In Sect. 2, we discuss how deep learning is leveraged on DeepFake algorithms for creating manipulated contents. Section 3 reviews a wide set of effective methods for DeepFake detection as well as their advantages and disadvantages. In Sect. 4, We discuss challenges, research trends, and directions on DeepFake generation and detection domains. Finally, Sect. 5 concludes the study.

2 DeepFake Generation

DeepFake contents get attraction because of the availability of robust and powerful set of DeepFake applications to a wide range of users. Such applications are capable to create a forged content by a few number of clicks within a few sec onds. Because of their rapid popularity, a large number of researches have been performed on Deep-Fake generation in recent days. In this section, we discuss about various DeepFake generation approaches and the datasets.

2.1 Generation Approaches

Most of the DeepFake related works have been done leveraging deep learning techniques. From the state-of-the-art literatures, DeepFake generation through facial image manipulation can be classified into four methodological categories based on the way and extent of manipulation: complete face synthesis, identity swap, attribute manipulation, and Face Reenactment. We provide a detailed discussion in the following sections. We discuss all other methods in a separate group—"Other DeepFake Generation Methods".

2.1.1 Complete Face Synthesis

Facial manipulation or editing techniques have been studied and developed considerably over the last few decades. It is also practised for generating DeepFake contents in recent days. Face synthesis generates photorealistic images of human faces that do not exist in real life. With the massive progression in Generative models, GAN [16], in the last few years, the research community has seen a significant amount of works associated with facial manipulation. GANs have been effectively used for generating photorealistic face images. Another variation of the generative deep learning model is variational autoencoder (VAE) [17] that also shows the potentiality in creating human face image.

Initially, these adversarial model starts generating realistic fake images from random vectors. The generative model tries to generate a more realistic image and fool the discriminative model in each iteration. On the contrary, the goal of the discriminative model is to verify the generated photo is either real or fake. Radford et al. proposed the deep convolutional generative adversarial network (DCGAN) [18], where the concept of both GAN and Convolutional Neural Network (CNN) has been utilized together to create a nonexisting human face. It is one of the initial works after the emergence of GAN in 2014. Liu et al. proposed VAE based CoGAN [19]. In COCO-GAN [20], the authors proposed a conditional GAN-based image generator that is capable of synthesizing images in a parallelizable fashion. However, Glow [21], a flow-based generative model, which is different from GAN's mechanism, is proposed by Kingma eet al. In this work, the authors used invertible 1×1 convolution for generating realistic DeepFake images.

Later in 2017, Wasserstein generative adversarial networks (WGAN) [22] has been proposed. The approach used in WGAN training is more stable than the previous method. Stability in GANs training was one of the primary issues in the first few years right after GAN's invention. WGAN minimizes this instability in GAN training. However, Gulrajani et al. [23] showed that due to the weight clipping operation, sometimes WGAN might fail to converge, thus might generate lousy images as output. In this paper, they provide an improved weight clipping approach to address the issue in WGAN training. BEGAN [24] is another work with the aim of improving WGAN. Karras et al. presented Progressive Growing GAN (PGGAN) [25] in 2017 with the focus on generating high-quality images. This is one of the pioneering works on generating high-quality images. The same author proposed StyleGAN [26] in 2019 that can automatically learn the highlevel attribute representation such as identity, pose to control different properties in generated images. StyleGAN2 [27], the extended version of the previous work, was presented in 2020.

2.1.2 Identity Swap

Identity swap is one of the most common face manipulation research techniques associated with DeepFakes. This approach includes replacing the human face in the target content (image or video) with another face in the source content. The traditional face swap process can be performed in three phases. First, the face is required to be detected in both source, and target content that can be done with face detection [28], or object detection model [29, 30]. The research community has seen numerous defensive [31] and offensive [32] applications with object detection techniques. After face or fa cial attributes detection in source and target content, the eyes, nose, eyebrows, mouth is replaced and adjusted and blended in term of lighting and color to minimize the difference between source and target content. In the third step, the adjusted candidates are ranked by the calculated distance over the overlapped region. However, this traditional face swap approach has limitations in generating very realistic face images as it offers static and rigid replacement. Different DL-based approaches have become very effective in realistic face-swapping with the super-progress in the Deep learning (DL) domain.

FaceSwap [33], and CycleGAN [34] are some of the very first works in this field. In FaceSwap, two sets of encoder–decoder combinations are used. The encoder part of the architecture is responsible for composing the latent feature of a face from the input image, and then the decoder part reconstructs the face. There are two parts of the training phase. In the first phase, Each encoder–decoder combination is trained with the source image. In the second phase of the training, the decoder gets trained with the target image. After successful training, the two decoders are substituted with each other. Consequently, the original encoder paired with the decoder of the target image is capable of constructing the target image with the facial features of the source image. The DeepFake generation (identity swap) procedure with pairs of encoder–decoder architecture is illustrated in Fig. 2.

The CycleGAN [34] solved the issue of the unavailability of paired training examples for image translation. FSGAN [35] is capable of face swapping and reenactment simultaneously with face reenactment and blending. Natsume et al. proposed two distinct VAE based RSGAN [36] to encode the latent representation of facial attributes. The recent work, FaceShifter [37] uses a two-phase scheme for high fidelity and occlusion-based face-swapping.

2.1.3 Attribute Manipulation

Face attribute manipulation is a process of modifying the specific region of the face in the target image. This process is very similar to face editing in most cases. Some examples of face manipulation are changing age, gender, hair color and style, disappearing hair (bald), and creating smiling faces from neutral faces, etc. Choi et al. applied the GAN concept in the image-to-image translation problem by presenting StarGAN [38] in 2018. In StarGAN, There is a single generator for translating images from a domain to multiple domains. However, StarGAN has a limitation as it can only generate a specific number of expressions. For Image translation, Chen et al. proposed HomointerpGAN [39], where during translation intermediate region between different domains is considered. The author suggested proper methods to select paths between two sample points in latent space to change particular image attributes.

Pumarola et al. mitigated this issue by proposing GANimation [40] where action unit (AU) annotation based GAN conditioning method is implemented. Later, the authors presented the improved version StarGANv2 [41] that can generate images with the highest visual quality. To develope this improved version, the authors design the model with encoder–decoder architecture where a random Gaussian noise vector fed to the generator. In terms of expression synthesis and attribute manipulation, StarGANv2 outperforms other works for its high scalability. To achieve better. He et al.

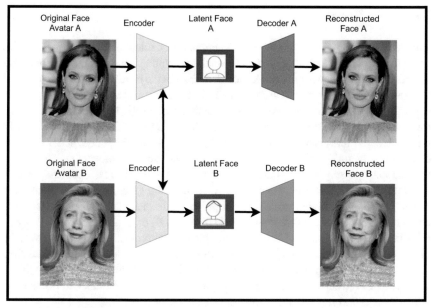

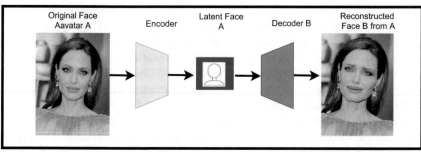

Fig. 2 A DeepFake creation model using two encoder–decoder combinations. In the training phases, the encoder–decoder set is used to learn the latent features of the input faces. While generation, two decoders are interchanged, such that latent face A is subjected to decoder B to generate the face A including the features of face B

introduced encoder–decoder-based AttGAN [42] where conditioned latent representation is used for more specific facial attribute editing as well as preserving other details of the face. One of the limitations of AttGAN is unwanted blurriness in generated face images. STGAN [43] is proposed by Liu et al. in 2019 as an improvement of AttGAN. In this work, the difference between target and source attribute is taken into consideration for more specific face attribute editing. However, STGAN shows poor performance in multiple attribute manipulation in the face image.

2.1.4 Face Reenactment

Face reenactment is a type of emerging DeepFake face manipulation technique, more precisely, which can be stated as a conditional face synthesis task for facial expression transfer. It refers to a process that replaces or transfers the facial expression of a person to another person. Face reenactment can be achieved by transferring the source actor's expression, gaze, pose, and mouth movement. Some of the most prominent works for real-time facial expression transfer have been done by Theis et al. [44, 45]. In Face2Face, the authors proposed methodologies to transfer the source person's facial expressions (actor), including facial gestures, head, and eye movement, to a video with another person (target person), maintaining the identity. Face2Face approach use deformation transfer between source face and target face, more specifically the mouth portion with higher priority for photo-realistic reenactment. For the tracking and reconstruction of the face identity (3D face model) of a source and target model, a commodity RGB-D sensor is used. After getting the required parameters from the 3D models, face expression is reenacted to the target face in each generated fake video frame. This approach is applicable in realtime standard RGB videos. In the same research group's successive work, FaceForensic++ [45] has been presented where NeuralTextures based learning approach has been utilized.

In the last few years, the research advancement in the development of GANs (Generative Adversarial Network) [16] is remarkable. GANs are comprised of two models, a generative model and a discriminative model, that are estimated through the adversarial process. Both models compete with each other to minimize their loss function in a fashion that can be interpreted as a minimax two-player game. GANs have been proven to be very effective for facial reenactment to generate realistic examples across a wide range of domains. The authors employ conditional GANs (cGANs) as a solution to image-to-image translation problems in their work [46] that is identical to the techniques from pix2pix software. Another work, Pix2pixHD [47], where authors proposed a successful high-resolution image generation approach using multi-scale cGANs with a perpetual loss. Wu et al. proposed ReenactGAN [48] to transfer both mouth and expression to the target face by mapping the source face boundary latent to the target face's boundary latent with a transformer. Eventually, reenacted target face is generated in a fake video with a target-specific decoder.

Zhang et al. [49] proposed a one-shot approach to generated reenacted faces using only a single source image. The authors presented an auto-encoder-based model that can learn a latent representation of both the source and target face representation. A similar but recent one-shot face reenactment model, FaR-GAN [50] has been proposed by Hao et al.

2.1.5 Other Generation Methods

Other than previously discussed approaches, some works with different methodologies might be classified as DeepFake generation approaches. This category includes lip-syncing, inpainting, style transfer, super-resolution, etc.

Lip-syncing DeepFake video generation approach produces a video of a target person in a fashion that the mouth and lip movement in the output video is synchronized with a source audio input. This synchronized mouth region movement makes the generated video realistic. Fan et al. introduced a deep bidirectional Long short-term memory (LSTM) based approach [51] for audio/visual modeling to develop a photo-real talking head system. LSTM is a subset of recurrent neural network (RNN) architecture that can model sequential data where longterm dependencies need to be considered. LSTM is widely adopted in prediction from health data [52], natural language processing (NLP) [53], next video frame prediction [54] etc. LSTM model with other DL models can learn and predict the lip movement from the input source audio file.

The image inpainting approach involves reconstructing the missing or incomplete part of images or videos. Yu et al. presented a well-known work, ContextAtten [55] in 2018 in image inpainting. The most common issue in the previous image inpainting is distorted structures or blurry texture in the manipulated image. Later, this issue is addressed in ContextAtten. SC-FEGAN [56] by Jo et al. is an image editing work focusing on utilizing a relatively free-form user input in terms of color and shape.

Single image super-resolution (SISR) task might be considered as a variation of DeepFake generation. Dai et al. proposed the Second-order attention network (SAN) [57] for more effective feature expression and correlation learning. In this work, the authors focus on feature correlation instead of using a model with deep architecture. Karnewar et al. proposed MSGGAN [58] for high-resolution image synthesis with a goal of achieving well convergence on a variety of image datasets. For increasing convergence stability, The authors allow the gradients-flow from the discriminator to the generator at various scales.

One of the most common recognizable factors in DeepFake images is artifacts in their frequency domain. Some of the DeepFake research community try to eliminate traceable artifacts by modifying the generation procedure. SDGAN [59], WUCGAN [60] are examples of such works. In WUCGAN, a spectral regularization has been used to overcome the GANs' inability to produce real image spectral distribution due to the up-sampling method.

In Table 1, we provide summaries of the DeepFake generation works mentioned above.

2.2 DeepFake Generation Dataset

A proper dataset plays a pivotal role in the deep learning model's performance. For DeepFake generation, two major types of datasets are used for different purposes: real dataset and synthesized dataset. In most cases, real datasets are used for DeepFake generation, whereas synthesized or fake datasets are used for DeepFake detection. Discussion on popular real datasets in the provided section below.

Yi et al. presented CASIA-WebFace [63] in 2014. The dataset includes 10,575 subjects and 494,414 images. CelebA [64] dataset was introduced by Liu et al.. This

Table 1 Summary of DeepFake Generation approaches. We list 26 recent approaches published in peer-reviewed journals and conferences in the table

Works	Methods	Elo rating	Datasets	Multimedia	
				Image	Video
Kingma and Dhariwal [21]	Glow	1511	CIFAR-10, ImageNet, LSUN	✓	
Jo and Park [56]	SC-FEGAN	1489	CelebA-HQ	✓	
Radford et al. [18]	SAN	1487	LSUN, Imagenet, Faces, CIFAR-10, SVHN, MNIST	✓	
Gulrajani et al. [23]	WGAN-GP	1435	LSUN, Google Billion Word, Swiss Roll	✓	
Yu et al. [61]	ContextAtten	1430	CelebA, CelebA-HQ, ImageNet, Places2, DTD	✓	
He et al. [42]	AttGAN	1426	CelebA, LFW	✓	
Lin et al. [20]	CocoGan	1400	CelebA, CelebA-HQ, LSUN, Matterport3D	✓	
Durall et al. [60]	WUCGAN	1400	FaceForensics++, CelebA, Faces, faces-HQ	✓	✓
Zhu et al. [34]	CycleGAN	1400	Cityscapes, CMP Facade, UT Zappos	✓	
Pumarola et al [40]	GANimation	1390	EmotioNet, RaFD	✓	
Choi et al. [38]	StarGAN	1387	CelebA, RaFD	✓	
Karnewar and Wang [58]	MSGGAN	1385	CelebA-HQ, CIFAR-10, OF, LSUN, FFHQ	✓	
Mao et al. [62]	DCGAN	1337	HWDB1.0, LSUN, MNIST	✓	
Chen et al. [39]	HomointerpGAN	1335	RaFD, CelebA	✓	
Karras et al. [25]	PGGAN	1336	CelebA, LSUN, CIFAR10	✓	
Berthelot et al. [24]	BEGAN	1291	CelebA	✓	
Natsume et al. [36]	RSGAN	N/A	CelebA	✓	
Hao et al. [50]	FaR-GAN	N/A	VoxCeleb1	✓	
Jung and Keuper [59]	SDGAN	N/A	FFHQ	✓	

(continued)

Table 1 (continued)

Works	Methods	Elo rating	Datasets	Multimedia	
				Image	Video
Liu et al. [43]	STGAN	N/A	FFHQ	✓	
Wu et al. [48]	ReenactGAN	1400	CelebV, WFLW		✓
Choi et al. [41]	StarGANV2	1400	CelebAHQ, AFHQ	✓	
Karraet al. [26]	StyleGAN	N/A	FFHQ	✓	
Karras et al. [27]	StyleGAN2	1416	FFHQ	✓	
Nirkin et al. [35]	FSGAN	1400	IJB-C, VGGFace2, Figaro, Forensics++, CelebA	✓	✓
Liu and Tuzel [19]	CoGAN	N/A	MNIST, USPS, CelebA, RGBD, NYU	✓	

is a labeled version of CelebFaces [65]. In CelebA dataset, there are 10,000 subjects, a total of 200,000 images where each subject has twenty samples.

VGGFace [66] is a large dataset with 2.6 million images of 2,622 subjects. Microsoft Celeb (MS-Celeb-1 M) [67] is another large image dataset, particularly for face recognition. This dataset contains 10 million face images of 100 k identities. Cao et al. introduced VGGFace2 [68] in 2018, containing 3.31 million images of 9131 persons. Images are collected from the internet. This dataset has a wide variation in age, ethnicity, and pose. Flickr-Faces-HQ [26] was presented in 2019 by Karras et al. containing 70 k high-resolution images.

3 DeepFake Detection

It is obvious that DeepFakes can be tremendous threats ranging from a person to the whole world. To avoid the threats, an effective set of DeepFake detection approaches are required. Subsequently, an increase amount of research is performed for developing various approaches to detect the authenticity of a still image or video content. In the past, a manipulated content was determined manually by analyzing artifacts and inconsistencies. In recent days, because of leveraging deep learning, complex and discriminative features are extracted to detect fake contents.

3.1   Detection Approaches

In this section, we review recent studies on various DeepFake detection works, based on their methodologies and extracted attributes.

3.1.1   Forensics-Based Detection

Recent forensics-based detection studies analyze pixel-level disparity. In addition, they provide explainable detection mechanism to determine authenticity however, these works undergo robustness issues when the manipulated contents are generated by simple transformations. Li et al. observed that the disparities between manipulated and real faces are revealed in the color components [69]. The authors proposed training a oneclass classifier on real face data based on considering the disparities in the color components to detect the unknown GANs. However, they did not clarify the performance of their approach against perturbation attacks such as image transformations.

In [70], Koopman et al. proposed to detect fake videos based on the unique noise pattern in the videos that is caused by the camera sensor. Rather than considering noise, the authors of [71] observed that DeepFakes usually contain inconsistent or unusual head poses with respect to expression. Hence, their work monitors facial landmarks to calculate the differences in head pose between manipulated and genuine video frames. They use the difference in estimated head pose data as a feature vector to train an SVM based classifier to predict original and DeepFakes. Unfortunately, the authors did not clarify the effectiveness of their works [70, 71] in detecting high quality DeepFakes.

In [72], Wang et al. leveraged the local motion features captured from real videos to identify the inconsistency of forged videos. They emphasized on the lowlevel features that are not feasible to be deployed in the wild where DeepFakes suffer known and unknown degradations.

Demir et al. [73] focused on synthetic eyes construction in deep fake videos. They generated features from eye and gaze data to train their model and compare it with complex state-of-the-art CNNs (VGG19, Inception, Xception, ResNet, and DenseNet). They claimed fake detection accuracy is around 6.5% higher than those of complex architectures without using eye or gaze information. However, considering only gaze data for detecting a synthesized content does not indicate improvement in generalization. Therefore, it is not clear that how the model will perform in detecting unseen adversary.

3.1.2   Deep Neural Network-(DNN) Based Detection

For DeepFakes detection in images and videos, neural network models with deep architecture outperform classical/hand-crafted approaches. DNN based models are capable of learning meaningful features from available data for effective forgery detection. Guera et al. [74] proposed a DeepFake video detection framework model combined with an RNN and CNN architecture to detect forged part from the input video. However, the limitation of the work is its incapability to handle videos for more than 2 s.

Nataraj et al. [75] presented a CCN model-based approach to calculate pixel co-occurrence matrices from the input image to detect image manipulation. Nguyen et al. [76] proposed a robust DeepFake video detector with multi-task CNN base architecture to identify and localize the manipulated portions from a video. An autoencoder and a decoder are used for the classification of manipulated content and sharing the extracted features for segmentation and shape reconstruction, respectively. However, the accuracy of this model declines for unseen examples that can be considered as a limitation of the work. To address this accuracy degradation-related limitation, a Forensic Transfer (FT) based CNN approach [77] for DeepFake detection was proposed by Stehouwer et al.

Marra et al. [78] presented an approach based on incremental learning for GAN-generated fake image detection. This work focuses on the classification and detection of a new type of GAN-generated images with high accuracy. This approach is more generalized and robust against detecting unseen GAN-generated examples. However, This procedure still needs some information about the new GAN architecture, which affects the practicality of the work.

3.1.3   GAN Redesign-Based Detection

In lieu of considering only audiovisual artifacts, a few number of research criticize the design limitation of existing GAN-based approaches and highlight to redesign GAN by including new artifacts.

In [93], McCloskey and Albright investigated the traditional architecture of the generator function and observed that the internal values of the generator are normalized. They claimed the normalization technique limits the frequency of the saturated pixels and makes it difficult to calculate the occurrences of saturated and underexposed pixels. They suggested to use their proposed approach as a complementary to other approaches that detect visual artifacts in the manipulated contents.

Zhang et al. investigated how the generalization ability of the existing detectors are impacted due to the existing upsampling design related artifacts [92]. They also noted that upsampling design is generic in GAN pipelines. Hence, they proposed a new signal processing analysis and redesigned the classifier accordingly. In addition, they developed a simulator framework, AutoGAN that simulates the common generation pipeline shared by a large class of popular GAN models [92]. AutoGAN simulates the GAN generation pipeline and generates (simulated) fake images that can be used in training any classifier without the burden of accessing pre-trained GANs.

In [94], Yu et al. proposed GAN fingerprint artifact for classifying the images and also determining the source of a target images. Although an insignificant amount of differences yield a distinct fingerprint, the fingerprints is vulnerable (i.e., tempering) to perturbation attacks such as image transformation, blur, JPEG compression, and so on.

3.1.4 Visual and Audio Inconsistency-Based Detection

Mittal et al. [85] work has addressed the essentiality of multimodal approach for DeepFake detection. They proposed an approach combining two modalities: the audio(speech) and video (face) to extract emotional features from both modalities to detect any kinds of counterfeit in the input video. This approach won't work if multiple persons are present in one video.

In [95], the authors tried to buck forgery in the realm of videos and images, by inspecting/traceback the source/mechanism of a given DeepFake image. ML tools are not always enough to combat this kind of problem. In addition, current robust DeepFake detection systems are vulnerable to adversarial images/videos. For these reasons, instead of building a robust DeepFake image/video detection system, it is more effective and scalable to find the associated generative model. With the help of a trusted third party, we can restrict/limit the malicious purpose of usage of this model. But deniability, misattribution to the original developer still a problem of the attribution approach. In easy terms, we can mention this attribution process as *Traitor Tracing*. This system will enforce accountability among model developers.

Along side with DeepFake video generation, audio spoofing is another way of character assassination of a public figure. Chintha et al. [91] addressed this problem by finding inconsistencies in audio and video modalities. To this end, the authors leveraged XceptionNet architecture for facial feature extraction and stacked convolutional layers to generate audio embedding features. Our analysis suggests that the combination of spoof audio and fake video detection is prone to achieve better generalization that indicates robustness in detecting unknown adversaries.

3.1.5 Other Notable Detectors

Rashmiranjan et al. (2021) in [96], investigated a technique involving Euler video magnification (EVM) process extracting features using three techniques (SSIM, LSTM, Heart Rate Estimation) to train models to classify counterfeit and unaltered videos. This technique uses spatial decomposition and temporal filtering on video data to highlight and magnify hidden features such as pulse of skin or subtle motions.

Fernandes et al. applied similar technique in [97], where they used EVM in color-based photoplethysmography (PPG) to identify blood volume fluctuations by shining light of certain wavelength onto the skin and measuring changes in light assimilation of the oxygenated blood which is in turn measures the heart rate. On the other hand, in this work, the authors applied both the EVM based color and movement amplification on videos to distinguish between original and fake videos. The results using SSIM technique when used a range of standard machine learning shows below 82% accuracy achieved by the best performing submission to the DFDC while the results using LSTM technique establishes apparent setback to the idea of using EVM for DeepFake detection. Overall, even though the color and spatial aspects of EVM were tested as possibilities for a number of classification models, the authors used accuracy as a metric though it is not known to be great metric for evaluation when imbalanced datasets are used which can be improved.

Hussain et al. [98] and Carlini et al. [99] discussed the vulnerability of current DeepFake detectors in light of both the whitebox and blackbox attacking approach. Carlini et al. [99] also demonstrated that a novice attacker can effectively conduct a blackbox attack without having any information regarding classifier and can reduce classifier's AUC to 0.22.

DeepTag [100] is another digital watermarking-based proactive system to combat DeepFake problems. This system finds the source of a DeepFaked image with an embedded message associated with the original image. This system works better against the dynamic image transformation and reconstruction of images by the Deep-Fake process. By blocking the confirmed DeepFake, this system also helps to stop spreading misinformation on the different social media platforms. According to their approach, the embedded message has to avoid the manipulated region. Even though the authors addressed this problem, they did not provide any solution on this.

DeepFake detection becomes more challenging when multiple faces are observed in a video frame. Charitidis et al. [101] tried to solve this problem with the improve-ment of preprocessing step. They pruned a cluster of facial data that carries less significance. This approach makes DeepFake detection process fast and it can be used on top of any existing DeepFake detection system. This approach still has one problem, their preprocessing approach can discard less prevalent but significant data from the datasets.

Table 2 shows the summary of the aforementioned DeepFake detection works and corresponding dataset information.

3.2 DeepFake Detection Dataset

The DeepFake Detection Challenge (DFDC) Preview Dataset

In [15], Dolhansky et al. introduced a preview of DFDC dataset containing 5,000 videos that featured two facial modification algorithms where the actors were in agreement to use and manipulate their likeliness. To ensure visual variability, diver-sity in several axes (gender, skin tone, age) and arbitrary background was considered. A reference performance baseline was provided in terms of specific metrics that was defined and tested on two existing models for detecting DeepFakes. The initial base-line consists of the performance check of three simple detection models. The first model is trained to detect low-level image and the other two models were trained on the FaceForensics++ dataset [102] and evaluated as implemented in [45]. All

Table 2 We present Summary of existing top-25 notable DeepFake detection work published in peer-reviewed journals and conferences in the table. We report only the highest achieved resultant metrics. ACC, PRE, AUC, and EER denotes accuracy, precision, area under the curve, and equal error rate, respectively

Works	Methods	Performance	Datasets	Multimedia	
				Image	Video
Li et al. [69]	One-class	ACC: 0.98	Self-built	✓	
Khalid and Woo [79]	VAE	ACC: 0.98	FF++		✓
Ciftci et al. [80]	CNN	ACC: 0.96	FF, FF++, Celeb-DF		✓
Li et al. [81]	S-MIL	ACC: 0.83	FF++, Celeb-DF, DFDC		✓
Masi et al. [82]	RNN	AUC: 0.99	FF++, Celeb-DF, DFDC		✓
Feng et al. [83]	CNN	AUC: 0.99	UADFV, Celeb-DF, FF++		✓
Dang et al. [84]	CNN	ACC: 0.98	Celeb-DF, UADFV, DFFD	✓	
Mittal et al. [85]	DNN	AUC: 0.96	TIMIT, DFDC		✓
Chai et al. [86]	CNN	PRE: 1.0	FF++	✓	✓
Li et al. [87]	HRnet	ACC: 0.95	UADFV, Celeb-DF, FF++		✓
Tarasiou and Zafeiriou [88]	CNN	ACC: 0.98	FF++		✓
Nguyen et al. [76]	CNN	ACC: 0.93	FF++		✓
Afchar et al. [89]	CNN	ACC: 0.98	FF++		✓
Sabir et al. [90]	RNN	ACC: 0.99	FF++		✓
Wang et al. [72]	CNN	PRE: 1.0	FF++	✓	✓
Güera and Delp [74]	RNN	ACC: 0.97	Self-built		✓
Koopman et al. [70]	N/A	N/A	Self-built		✓
Nataraj et al. [75]	CNN	ACC: 0.99	Self-built	✓	
Demir and Ciftci [73]	CNN	ACC: 0.89	FF++		✓
Chintha et al. [91]	CNN	EER 0.13	FF++, Celeb-DF		✓
Cozzolino et al. [77]	CNN	ACC 1.0	Self-built	✓	
Zhang et al [92]	CNN	ACC: 1.0	Self-built	✓	
McCloskey and Albright [93]	SVM	AUC: 0.91	Self-built	✓	
Marra et al. [78]	N/A	N/A	Self-built	✓	

(continued)

Table 2 (continued)

Works	Methods	Performance	Datasets	Multimedia	
				Image	Video
Yu et al. [94]	CNN	ACC: 0.99	Self-built	✓	

performances of these three sample detection models were analyzed using precision, recall, and the logarithmic scale of weighted precision to detect half, most, or nearly-all DeepFakes.

The Celeb-DF Dataset

At least until the year 2019, DeepFake datasets included low visual quality and had little to no resemblance to DeepFake videos found online. The work presented in [103] has constructed a large scale DeepFake video dataset called *Celeb-DF* that includes a total of 5,639 high-quality DeepFake videos, corresponding to more that 2 million frames from publicly available YouTube video clips of 59 celebrities of diverse genders, ages, and ethnic groups using improved synthesis process. The video quality in *Celeb-DF* with very few notable visual artifacts have significant differences with then available DeepFake videos available online that included low-quality synthesized faces, visible splicing boundaries, color mismatch, and inconsistencies in synthesized face orientation etc. The overall quality of the videos were enhanced in terms of improving low resolution of synthesized faces, color mismatch, inaccurate face masks, and temporal flickering. The authors also presented a comprehensive evaluation with 9 DeepFake detection methods and datasets considered making the most comprehensive study of DeepFake detection available by then. Overall, this *Celeb-DF* dataset has helped lowered the gap in the video quality between the actual and DeepFake datasets that can be found online with a possibility of enlarging *Celeb-DF* and further enhancing the visual quality including the running efficiency.

The FaceForensics++ Dataset

Rossler et al. in [102] generated a large-scale dataset with an automated benchmark based on classical computer-graphics and learning-based based methods such as DeepFakes [104], FaceSwap [105], NeuralTextures [106], and Face2Face [107]. This benchmark contains a hidden test set of an order magnitude larger than comparable, publicly available dataset including 1.8 million manipulated images extracted from 1,000 real videos and target ground truth to enable supervised learning. The authors conducted a thorough study of data-driven forgery detectors and showed that the use of domain specific information in conjunction with a XceptionNet classifier improves the detection with an unprecedented accuracy. This work also presented ways to automatically detect any forms of facial identity and facial expression manipulations with an automated benchmark consisting with random compression and dimensions.

The UADFV Dataset

Up until the revelation of the popular software "DeepFake"that used generative adversary networks (GANs), since any form of manipulation of videos/images involved huge time consumption for editing operations, realistic high quality fake videos were not widespread. Due to this software, ample of high-quality fake video flooded the internet and thus detecting such videos became important. In [108], the authors used 50 YouTube videos that lasted 30 s each representing one individual with at least one blinking occurred, to form the Eye Blinking Video (EBV) dataset. The left and right states of each frame were annotated with a user-friendly annotation tool. The training dataset that was used in this work is CEW [109] which includes 1,793 images of closed eyes and 1,232 images of open eyes to train front-end CNN model, 40 videos as training set for the overall Long-term Recurrent Convolutional Networks (LRCN) model [110] and 10 videos as the testing set. This LRCN method was shown to exhibit best performance 0.99 compared to other methods such as Eye Aspect Ratio (EAR) [111] with performance 0.79 and CNN with 0.98.

4 Challenges and Opportunities for Future Research Direction

We review over 60 articles published either in peer-reviewed journals and conferences or posted on arXiv regarding DeepFake generation and detection. In this section, we describe our observations including challenges, limitations, and new research scopes, after reviewing the studied articles. This will contribute on the future research in creating more realistic and detection-evasive DeepFake, and also sophisticated DeepFake detection model.

4.1 Findings in DeepFake Generation

– We notice low resolution and poor quality in the output image irrespective to any existing DeepFake generation work. Currently, generation of highresolution and sharp images is difficult since such image makes the job easier to differentiate it from training images [112]. A deeper analysis claims that this causes spike in the gradient problem and affects training stability [113]. To mitigate the challenge, PGGAN is proposed to grow generator and discriminator progressively. It starts from low-resolution images and gradually, adds new layers for higher-resolution (i.e., details) as the training progresses [113]. However, PGGAN is still in premature stage and capable of generating only (1024 × 1024) size images.
– The attribute manipulation methods are limited as these can only change the properties followed by the training set. Therefore, such an attribute manipulation

method is needed that could capture attributes are independent attributes to the training set.

- In most of the cases, identity swap and expression swap do not consider the continuity of the videos. They neither consider gesture nor physiological signals such as eye blink, breathing frequency, heart beat, and so on.
- The fake datasets are expanded only considering the diversity of the contentre-lated factors such as age, gender, background, and so on. Our investigation conform factors such as added noise (quality degradation), Gaussian blue, JPEG compression, contrast change, and so on can enhance the diversity in the dataset. DeeperForensics-1.0 [114] dataset offers image-level degradations but it is added artificially by post-processing. We expect natural image/video-level degradations (i.e., over/underexposed photos, bit-rate variations, choices of codec) in the future generation of the dataset.

4.2 Findings in DeepFake Detection

- Most of the existing works generate image dataset to evaluate the effectiveness of their approaches leveraging various GANs. A large portion of the works do not unveil the details about the datasets that used in evaluation. Hence, the quality of the generated forged images still remains unknown. On the contrary, these works claim their competitive results in detecting various synthesized images built on their own. We emphasize on the development of public GAN-synthesized fake image dataset.
- For the sake of performance evaluation, existing works implement simple base-lines (e.g., vanilla DNN-based methods) and compare it with their works rather than considering state-of-the-art ones. Claiming the superiority of their works by comparing with naïve approaches indicate biased evaluation. We expect that future works should be comparable to state-of-the-art works so that we can understand effectiveness of the proposed work.
- The aim of DeepFake detection research is to develop more robust and gener-alized approaches. Subsequently, the research community is trying their level best to come up with effective approaches. However, the recent works are simply evaluated on simple DeepFake video datasets, such as FaceForensics++ (FF++). We emphasize that future works should focus more on challenging datasets for acceptable performance evaluation.
- Almost all of the existing studies report their experimental results by merely considering the detection accuracy without reporting other popular metrics such as precision, recall, and the relation with the quality of DeepFakes. To conduct an acceptable performance evaluation, a comprehensive experimental result set is mandatory. A robust set of experiments should contain the result of various effective metrics. Currently, there does not exist any metric that can measure DeepFake quality. We hope, in future, the researchers would come up with a new metric for measuring the quality of DeepFakes.

– Detecting the emerging unknown DeepFakes is crucial in today's world. Hence, developing a practical DeepFake detector that is deployable in the wild is a necessity. Towards developing an effective DeepFake detector, we observe a set of key factors that are: (1) advance generalization capabilities, robust against various attacks (e.g., iamge/video transformations, adversarial attacks), and presenting explainable DeepFake detection result. Unfortunately, in reviewing the recent DeepFake detection articles, we find that the researchers simply ignore to evaluate the capabilities of their works from the aforementioned perspectives.

5 Conclusion

Due to great progress on generative deep learning algorithm in recent years, nowadays it has become a real challenge to identify the authenticity of any visual content found online [115]. The aim of creating the synthesized contents is either for malicious intent or just for fun. To resist any unexpected scenarios such as creating a manipulated content of important persons (e.g., political leaders, celebrities) or generate synthesized contents for a useful purpose, the current DeepFake research community needs to consider the existing published articles both in DeepFake generation and detection to plan for extensive research efforts in the future. In light on this and to make our understanding better, in this current work, our investigation shows that in recent years deep learning research community have been trying to solve two large research domains including DeepFake detection and generation related to Deep-Fake image and video contents. We have shed light on these domains by discussing stateof-the-art research works. We also try to depict how the research community shifts their attention from feature-based DeepFake detection to feature agnostic and policy-based approaches to combat evasion of DeepFake detection. We also provide comprehensive descriptions of different prominent datasets to facilitate researchers to determine their next research direction.

References

1. Goodfellow I, Pouget-Abadie J, Mirza M, Xu B, Warde-Farley D, Ozair S, Courville A, Bengio Y, Generative adversarial nets
2. Deepfakes porn has serious consequences. https://www.bbc.com/news/ technology-42912529. Accessed 1 Apr 2021
3. Deepfake Porn Nearly Ruined My Life. https://www.elle.com/uk/ life-and-culture/a30748079/deepfake-porn. Accessed 1 Apr 2021
4. deepfake examples that terrified and amused the internet. https:// www.creativebloq.com/features/deepfake-examples. Accessed 1 Apr 2021
5. Kaliyar RK, Goswami A, Narang P (2020) Deepfake: improving fake news detection using tensor decomposition-based deep neural network. J Supercomput 1–23
6. A guy made a deepfake app to turn photos of women into nudes. It didn't go well. https://www. vox.com/2019/6/27/18761639/ai-deepfake-deepnude-app-nude-women-porn. Accessed 1 Apr 2021

7. Chinese deepfake app Zao sparks privacy row after going viral. https://www.theguardian.com/technology/2019/sep/02/chinese-face-swap-app-zao-triggers-privacy-fears-viral. Accessed 1 Apr 2021
8. Lyu S (2020) Deepfake detection: current challenges and next steps. In: 2020 IEEE international conference on multimedia & expo workshops (ICMEW), pp 1–6
9. Guarnera L, Giudice O, Nastasi C, Battiato S (2020) Preliminary forensics analysis of deepfake images. In: Proceedings of international annual conference (AEIT), pp 1–6
10. Juefei-Xu F, Wang R, Huang Y, Guo Q, Ma L, Liu Y (2021) Countering malicious deepfakes: survey, battleground, and horizon. arXiv preprint arXiv:2103.00218
11. Jafar MT, Ababneh M, Al-Zoube M, Elhassan A (2020) Forensics and analysis of deepfake videos. In: Proceedings of the 11th international conference on information and communication systems (ICICS), pp 053–058
12. Trinh L, Tsang M, Rambhatla S, Liu Y (2020) Inter-pretable deepfake detection via dynamic prototypes. arXiv preprint arXiv:2006.15473
13. S.3805—malicious deep fake prohibition act of 2018. https://www.congress.gov/bill/115th-congress/senate-bill/3805. Accessed 1 Apr 2021
14. Help us shape our approach to synthetic and manipulated media. https://blog.twitter.com/en_us/topics/company/2019/synthetic_manipulated_media_policy_feedback.html. Accessed 1 Apr 2021
15. Dolhansky B, Howes R, Pflaum B, Baram N, Ferrer CC (2019) The deepfake detection challenge (dfdc) preview dataset. arXiv preprint arXiv:1910.08854
16. Goodfellow IJ, Pouget-Abadie J, Mirza M, Xu B, Warde-Farley D, Ozair S, Courville A, Bengio Y (2014) Generative adversarial networks
17. Diederik P Kingma and Max Welling. Auto-encoding variational bayes. arXiv preprint arXiv:1312.6114
18. Radford A, Metz L, Chintala S (2015) Unsupervised representation learning with deep convolutional generative adversarial networks.arXiv preprint arXiv:1511.06434
19. Liu MY, Tuzel O (2016)Coupled generative adversarial networks. arXiv preprint arXiv:1606.07536
20. Lin CH, Chang CC, Chen YS, Juan DC, Wei W, Chen HT (2019) Coco-gan: generation by parts via conditional coordinating.In Proceedings of the IEEE/CVF international conference on computer vision, pp 4512–4521
21. Kingma DP, Dhariwal P (2018) Glow: generative flow with invertible 1x1 convolutions. arXiv preprint arXiv:1807.03039
22. Arjovsky M, Chintala S, Bottou L (2017) Wasserstein generative adversarial networks. In: Proceedings of international conference on machine learning. PMLR, pp 214–223
23. Gulrajani I, Ahmed F, Arjovsky M, Dumoulin V, Courville A (2017) Improved training of wasserstein gans. arXiv preprint arXiv:1704.00028
24. Berthelot D, Schumm T, Metz L (2017) Began: boundary equi librium generative adversarial networks. arXiv preprint arXiv:1703.10717
25. Karras T, Aila T, Laine S, Lehtinen J (2017) Progressive growing of gans for improved quality, stability, and variation. arXiv preprint arXiv:1710.10196
26. Karras T, Laine S, Aila T (2019) A style-based generator architecture for generative adversarial networks. In: Proceedings of the IEEE/CVF conference on computer vision and pattern recognition, pp 4401–4410
27. Karras T, Laine S, Aittala M, Hellsten J, Lehtinen J, Aila T (2020) Analyzing and improving the image quality of style-gan. In Proceedings of the IEEE/CVF conference on computer vision and pattern recognition, pp 8110–8119
28. Zhang K, Zhang Z, Li Z, Qiao Y (2016) Joint face detection and alignment using multitask cascaded convolutional networks. IEEE Signal Process Lett 23(10):1499–1503
29. Liu W, Anguelov D, Erhan D, Szegedy C, Reed S, Fu C-Y, Berg AC (2016) Ssd: single shot multi- box detector. In: European conference on computer vision. Springer, pp 21–37
30. Redmon J, Divvala S, Girshick R, Farhadi A (2016) You only look once: unified, real-time object detection. In: Proceedings of the IEEE conference on computer vision and pattern recognition, pp 779–788

31. Akcay S, Kundegorski ME, Willcocks CG, Breckon TP (2018) Using deep convolutional neural network architectures for object classification and detection within x-ray baggage security imagery. IEEE Trans Inf Forens Secur 13(9):2203–2215
32. Hossen MI, Tu Y, Rabby MF, Islam MN, Cao H, Hei X (2020) An object detection based solver for google's image recaptcha v2. In: 23rd international symposium on research in attacks, intrusions and defenses (RAID 2020), pp 269–284
33. FaceSwap (2016). https://github.com/deepfakes/faceswap
34. Zhu J-Y, Park T, Isola P, Efros AA (2017) Unpaired image-to-image translation using cycle-consistent adversarial networks. In: Proceedings of the IEEE international conference on computer vision, pp 2223–2232
35. Nirkin Y, Keller Y, Hassner T (2019) Fsgan: subject agnostic face swapping and reenactment. In: Proceedings of the IEEE/CVF international conference on computer vision, pp 7184–7193
36. Natsume R, Yatagawa T, Morishima S (2018) Rsgan: face swapping and editing using face and hair representation in latent spaces. arXiv preprint arXiv:1804.03447
37. Li L, Bao J, Yang H, Chen D, Wen F (2019) Faceshifter: towards high fidelity and occlusion aware face swapping. arXiv preprint arXiv:1912.13457
38. Choi Y, Choi M, Kim M, Ha J-W, Kim S, Choo J (2018) Stargan: unified generative adversarial networks for multi-domain image-to-image translation. In: Proceedings of the IEEE conference on computer vision and pattern recognition, pp 8789–8797
39. Chen Y-C, Xu X, Tian Z, Jia J (2019) Homomorphic latent space interpolation for unpaired image-to-image translation. In: Proceedings of the IEEE/CVF conference on computer vision and pattern recognition, pp 2408–2416
40. Pumarola A, Agudo A, Martinez AM, Sanfeliu A, Moreno-Noguer F (2018) Ganimation: anatomically-aware facial an- imation from a single image. In: Proceedings of the European conference on computer vision (ECCV), pp 818–833
41. Choi Y, Uh Y, Yoo J, Ha J-W (2020) Stargan v2: diverse image synthesis for multiple domains. In: Proceedings of the IEEE/CVF conference on computer vision and pattern recognition, pp 8188–8197
42. He Z, Zuo W, Kan M, Shan S, Chen X (2019) Attgan: facial attribute editing by only changing what you want. IEEE Trans Image Process 28(11):5464–5478
43. Liu M, Ding Y, Xia M, Liu X, Ding E, Zuo W, Wen S (2019) Stgan: a unified selective transfer network for arbitrary image attribute editing. In: Proceedings of the IEEE/CVF conference on computer vision and pattern recognition, pp 3673–3682
44. Thies J, Zollh¨ofer M, Stamminger M, Theobalt C, Nießner M (2019) Face2face: real-time face capture and reenactment of rgb videos. abs/2007.14808
45. R¨ossler A, Cozzolino D, Verdoliva L, Riess C, Thies J, Nießner M (2019) FaceForensics++: learning to detect manipulated facial images. In: International conference on computer vi son (ICCV)
46. Isola P, Zhu J-Y, Zhou T, Efros AA (2018) Image-to-image translation with conditional adversarial networks
47. Wang T-C, Liu M-Y, Zhu J-Y, Tao A, Kautz J, Catanzaro B (2018) High-resolution image synthesis and semantic manipulation with conditional gans
48. Wu W, Zhang Y, Li C, Qian C, Loy CC (2018) Reenactgan: learning to reenact faces via boundary transfer. arXiv:1807.11079
49. Zhang Y, Zhang S, He Y, Li C, Loy CC, Liu Z (2019) One-shot face reenactment. arXiv:abs/1908.03251
50. Hao H, Baireddy S, Reibman A, Delp E (2020) Far-gan for one-shot face reenactment. arXiv: abs/2005.06402
51. Fan B, Wang L, Soong FK, Xie L (2015) Photo-real talking head with deep bidirectional lstm. In: 2015 IEEE international conference on acoustics, speech and signal processing (ICASSP). IEEE, pp 4884–4888
52. Rabby MF, Tu Y, Hossen MI, Lee I, Maida As, Hei X (2021) Stacked lstm based deep recurrent neural network with kalman smoothing for blood glucose prediction. BMC Med Inform Decis Mak 21(1):1–15

53. Zobaed S, Haque ME, Rabby MF, Amini Salehi M (2021) Senspick: sense picking for word sense disambiguation. In: 2021 IEEE 15th international conference on semantic computing (ICSC). IEEE, pp 318–324
54. Hosseini M, Maida AS, Hosseini M, Raju G (2020) Inception lstm for next-frame video prediction (student abstract). In: Proceedings of the AAAI conference on artificial intelligence, vol 34, pp 13809–13810
55. Yu J, Lin Z, Yang J, Shen X, Lu X, Huang TS (2018) Generative image inpainting with contextual attention. In: Proceedings of the IEEE conference on computer vision and pattern recognition (CVPR)
56. Jo Y, Park J (2019) Sc-fegan: face editing generative adversar ial network with user's sketch and color. In: Proceedings of the IEEE/CVF international conference on computer vision, pp 1745–1753
57. Dai T, Cai J, Zhang Y, Xia S-T, Zhang L (2019) Second-order attention network for single image super-resolution. In: Proceedings of the IEEE/CVF conference on computer vision and pattern recognition, pp 11065–11074
58. Karnewar A, Wang O (2020) Msg-gan: multi-scale gradients for generative adversarial networks. In: Proceedings of the IEEE/CVF conference on computer vision and pattern recognition, pp 7799–7808
59. Jung S, Keuper M (2020) Spectral distribution aware image generation. arXiv preprint arXiv: 2012.03110
60. Durall R, Keuper M, Keuper J (2020) Watch your up-convolution: CNN based generative deep neural networks are failing to reproduce spectral distributions. In: Proceedings of the IEEE/CVF conference on computer vision and pattern recognition, pp 7890–7899
61. Yu J, Lin Z, Yang J, Shen X, Lu X, Huang TS (2018) Generative image in painting with contextual attention. In: Proceedings of the IEEE conference on computer vision and pattern recognition, pp 5505–5514
62. Mao X, Li Q, Xie H, Lau RYK, Wang Z (2016) Multi-class generative adversarial networks with the l2 loss function 5:00102. arXiv preprint arXiv:1611.04076
63. Yi D, Lei Z, Liao S, Li SZ (2014) Learning face representation from scratch. arXiv preprint arXiv:1411.7923
64. Liu Z, Luo P, Wang X, Tang X (2015) Deep learning face attributes in the wild. In: Proceedings of the IEEE international conference on computer vision, pp 3730–3738
65. Sun Y, Wang X, Tang X (2013) Hybrid deep learning for face verification. In: Proceedings of the IEEE international conference on computer vision, pp 1489–1496
66. Parkhi OM, Vedaldi A, Zisserman A (2015) Deep face recognition
67. Guo Y, Zhang L, Hu Y, He X, Gao J (2016) Ms-celeb-1m: a dataset and benchmark for large-scale face recognition. In: European conference on computer vision. Springer, pp 87–102
68. Cao Q, Shen L, Xie W, Parkhi OM, Zisserman A (2018) Vggface2: a dataset for recognising faces across pose and age. In: 2018 13th IEEE international conference on automatic face & gesture recognition (FG 2018). IEEE, pp 67–74
69. Li H, Li B, Tan S, Huang J (2020) Identification of deep network generated images using disparities in color components. J Sign Process 174:107616
70. Koopman M, Rodriguez AM, Geradts Z (2018) De tection of deepfake video manipulation. In: Proceedings of the 20th Irish machine vision and image processing conference (IMVIP), pp 133–136
71. Yang X, Li X, Lyu S (2019) Exposing deep fakes using inconsistent head poses. In: Proceedings of international conference on acoustics, speech and signal processing (ICASSP). IEEE, pp 8261–8265
72. Wang S-Y, Wang O, Zhang R, Owens A, Efros AA (2020) CNN-generated images are surprisingly easy to spot... for now. In: Proceedings of the IEEE/CVF conference on computer vision and pattern recognition, pp 8695–8704
73. Demir I, Ciftci UA (2021) Where do deep fakes look? synthetic face detection via gaze tracking. arXiv preprint arXiv:2101.01165

74. Gu"era D, Delp EJ (2018) Deepfake video detection using recurrent neural networks. In: 2018 15th IEEE international conference on advanced video and signal based surveillance (AVSS). IEEE, pp 1–6
75. Nataraj L, Mohammed TM, Manjunath BS, Chandrasekaran S, Flenner A, Bappy JH, Roy-Chowdhury AK (2019) Detecting gan generated fake images using co- occurrence matrices. Electron Imag 2019(5):532
76. Nguyen HH, Fang F, Yamagishi J, Echizen I (2019) Multi- task learning for detecting and segmenting manipulated facial images and videos. In: 2019 IEEE 10th international conference on biometrics the ory, applications and systems (BTAS). IEEE, pp 1–8
77. Cozzolino D, Thies J, R"ossler A, Riess C, Nießner M, Verdoliva L (2018) Forensictransfer: weakly-supervised domain adaptation for forgery detection. arXiv preprint arXiv:1812.02510
78. Marra F, Saltori C, Boato G, Verdoliva L (2019) Incremental learning for the detection and classification of gan-generated images. In: 2019 IEEE international workshop on information forensics and security (WIFS). IEEE, pp 1–6
79. Khalid H, Woo SS (2020) Oc-fakedect: classifying deepfakes using one-class variational autoencoder. In: Proceedings of the IEEE/CVF conference on computer vision and pattern recognition workshops, pp 656–657
80. Ciftci UA, Demir I, Yin L (2020) Fakecatcher: detection of synthetic portrait videos using biological signals. IEEE Trans Pattern Anal Mach Intell
81. Li X, Lang Y, Chen Y, Mao X, Yuan He, Shuhui Wang, Hui Xue, and Quan Lu. Sharp multiple instance learning for deep- fake video detection. In: Proceedings of the 28th ACM international conference on multimedia, pp 1864–1872
82. Masi I, Killekar A, Mascarenhas RM, Gurudatt SP, AbdAlmageed W (2020) Two-branch recurrent network for isolating deepfakes in videos. In: European conference on computer vision. Springer, pp 667–684
83. Feng D, Lu X, Lin X (2020) Deep detection for face manipulation. In: International conference on neural information processing. Springer, pp 316–323
84. Dang H, Liu F, Stehouwer J, Liu X, Jain AK (2020) On the detection of digital face manipula-tion. In: Proceedings of the IEEE/CVF conference on computer vision and pattern recognition, pp 5781–5790
85. Mittal T, Bhattacharya U, Chandra R, Bera A, Manocha D (2020) Emotions don't lie: a deepfake detection method using audio-visual affective cues. arXiv preprint arXiv:2003.06711
86. Chai L, Bau D, Lim S-N, Isola P (2020) What makes fake images detectable? understanding properties that generalize. In: European conference on computer vision. Springer, pp 103–120
87. Li L, Bao J, Zhang T, Yang H, Chen D, Wen F, Guo B (2020) Face X-ray for more general face forgery detection. In: Proceedings of the IEEE/CVF conference on computer vision and pattern recognition, pp 5001–5010
88. Tarasiou M, Zafeiriou S (2020) Extracting deep local features to detect manipulated images of human faces. In 2020 IEEE international conference on image processing (ICIP). IEEE, pp 1821–1825
89. Afchar D, Nozick V, Yamagishi J, Echizen I (2018) Mesonet: a compact facial video forgery detection network. In: 2018 IEEE international workshop on information forensics and security (WIFS). IEEE, pp 1–7
90. Sabir E, Cheng J, Jaiswal A, AbdAlmageed W, Masi I, Natarajan P (2019) Recurrent convolutional strategies for face manipulation detection in videos. Interfaces (GUI) 3(1)
91. Chintha A, Thai B, Sohrawardi SJ, Bhatt K, Hickerson A, Wright M, Ptucha R (2020) Recur-rent convolutional structures for audio spoof and video deepfake detection. IEEE J Sel Top Signal Process 14(5):1024–1037
92. Zhang X, Karaman S, Chang S-F (2019) Detecting and simulating artifacts in gan fake images. In: 2019 IEEE international workshop on information forensics and security (WIFS). IEEE, pp 1–6
93. McCloskey S, Albright M (2019) Detecting gan-generated imagery using saturation cues. In: 2019 IEEE international conference on image processing (ICIP). IEEE, pp 4584–4588

94. Yu N, Davis LS, Fritz M (2019) Attributing fake images to gans: Learning and analyzing gan fingerprints. In: Proceedings of the IEEE/CVF international conference on computer vision, pp 7556–7566
95. Zhang B, Zhou JP, Shumailov I, Papernot N (2021) On attribution of deepfakes
96. Das R, Negi G, Smeaton AF (2020) Detecting deepfake videos using Euler video magnification. arXiv preprint arXiv:2101.11563
97. Fernandes S, Raj S, Ortiz E, Vintila I, Salter M, Urosevic G, Jha S (2019) Predicting heart rate variations of deepfake videos using neural ode. In: Proceedings of the IEEE/CVF international conference on computer vision workshops, pp 0–0
98. Hussain S, Neekhara P, Jere M, Koushanfar F, McAuley J (2021) Adversarial deepfakes: evaluating vulnerability of deepfake detectors to adversarial examples. In: Proceedings of the IEEE/CVF winter conference on applications of computer vision, pp 3348–3357
99. Carlini N, Farid H (2020) Evading deepfake-image detectors with white-and black-box attacks. In: Proceedings of the IEEE/CVF conference on computer vision and pattern recognition workshops, pp 658–659
100. Wang R, Juefei-Xu F, Guo Q, Huang Y, Ma L, Liu Y, Wang L (2020) Deeptag: Robust image tagging for deepfake provenance. arXiv preprint arXiv:2009.09869
101. Charitidis P, Kordopatis-Zilos G, Papadopoulos S, Kompatsiaris, Investigating the impact of pre-processing andprediction aggregation on the deepfake detection task
102. Rossler A, Cozzolino D, Verdoliva L, Riess C, Thies J, Nießner M (2019)Faceforensics++: learning to detect manipulated facial images. In Proceedings of the IEEE/CVF international conference on computer vision, pp 1–11
103. Li Y, Yang X, Sun P, Qi H, Lyu S (2020) Celeb-df:a large-scale challenging dataset for deepfake forensics. In: Proceedings of the IEEE/CVF conference on computer vision and pattern recognition, pp 3207–3216
104. Deepfakes github. https://github.com/deepfakes/faceswap. Accessed 04 Nov 2021
105. FaceSwap github. https://github.com/MarekKowalski/FaceSwap/. Accessed 04 Nov 2021
106. Thies J, Zollh¨ofer M, Nießner M (2019) Deferred neural rendering: image synthesis using neural textures. ACM Trans Graph (TOG) 38(4)1–12
107. Thies J, Zollhofer M, Stamminger M, Theobalt C, Nießner M (2016) Face2face: real-time face capture and reenactment of rgb videos. In: Proceedings of the IEEE conference on computer vision and pattern recognition, pp 2387–2395
108. Li Y, Chang M-C, Lyu S (2018) In ictu oculi: exposing ai generated fake face videos by detecting eye blinking. arXiv preprint arXiv:1806.02877
109. Song F, Tan X, Liu X, Chen S (2014) Eyes closeness detection from still images with multi-scale histograms of principal oriented gradients. Pattern Recogn 47(9):2825–2838
110. Donahue J, Hendricks LA, Guadarrama S, Rohrbach M, Venugopalan S, Saenko K, Darrell T (2015) Long-term recurrent convolutional networks for visual recognition and description. In: Proceedings of the IEEE conference on computer vision and pattern recognition, pp 2625–2634
111. Cech J, Soukupova T (2016) Real-time eye blink detection using facial landmarks. Cent. Mach. Perception, Dep. Cybern. Fac. Electr. Eng. Czech Tech. Univ. Prague 1–8
112. Odena A, Olah C, Shlens J (2017) Conditional image synthesis with auxiliary classifier gans. In: International conference on machine learning. PMLR, pp 2642–2651
113. Karras T, Aila T, Laine S, Lehtinen J (2018) Progressive growing of gans for improved quality, stability, and variation. In: Proceedings of international conference on learning representations
114. Jiang L, Li R, Wu W, Qian C, Loy CC (2020) Deeperforensics-1.0: a large-scale dataset for real-world face forgery detection. In: Proceedings of the IEEE/CVF conference on computer vision and pattern recognition, pp 2889–2898
115. Mirsky Y, Lee W (2021) The creation and detection of deepfakes: a survey. ACM Comput Surv (CSUR) 54(1):1–41

A Survey of Challenges Posed by the Dark Web

Brendan Staley and Reza Montasari

Abstract The dark nets' anonymity has provided the organised crime groups, terror groups, and paedophiles alike a place to communicate, recruit, purchase and disseminate illegal materials across the world instantly without much fear of retribution. As the exit nodes of the TOR browser can be set up by anyone across the globe, policing requires a significant degree of both skill and resources. Undercover operations within law enforcement agencies have been largely successful with paedophile rings being disrupted, as well as illicit cryptomarket places being monitored and shut down. However, this activity is incredibly time-consuming and costly. This chapters provides a discussion of some of the existing challenges that the darknet poses to law enforcement and other security agencies.

1 Introduction

What is commonly considered the Internet is essentially a linked network of systems. Any pages accessible through search engine indexes such as Google or Bing account for what is referred to as the "surface web". The surface web contains over 980 million websites but only accounts for roughly 4% of the entire Internet, which is why the Internet structure is often said to resemble that of an iceberg, as the majority remains "under the surface" and hidden. The remaining portion of the Internet is approximately equally split, between what is referred to as the "deep web" and the "darknet" [9]. The deep web contains all the data which is held on secure servers and requires specific addresses or login credentials such as academic/company user logins, Internet banking, and military purposes. The darknet was originally produced

B. Staley · R. Montasari (✉)
Hillary Rodham Clinton School of Law, Swansea University, Richard Price Building, Singleton Park, Swansea, UK
e-mail: Reza.Montasari@Swansea.ac.uk

B. Staley
e-mail: 827509@Swansea.ac.uk
URL: http://www.swansea.ac.uk

© The Author(s), under exclusive license to Springer Nature Switzerland AG 2021
R. Montasari and H. Jahankhani (eds.), *Artificial Intelligence in Cyber Security: Impact and Implications*, Advanced Sciences and Technologies for Security Applications,
https://doi.org/10.1007/978-3-030-88040-8_8

by the United States Naval Research Laboratory in order to provide a system in which agents and military personnel could communicate online without being identified and tracked [4]. The darknet or dark web is a network of global computers that use a cryptographic protocol to communicate, enabling users to conduct transactions and share data with anonymity. Some of the darknet community comprises of researchers, the curious, journalists, and whistle-blowers who share humour, knowledge, programming, and other educational programs. Although still controversially funded by the United States Government, the dark net's anonymity has captured the attention across the globe of those with untoward intentions. However, the controversy arises due to the proliferation and accessibility of drug dealing, hacking materials, stolen financial data, criminal groups, terrorists, and illegal pornography across the darknet. The darknet is accessible using TOR (The Onion Router), which is a free "circuit-based low-latency communication service" [10].

Although initially created for the US military, it now facilitates anonymous accessibility to web browsing and communication platforms. The anonymity allows the users significantly more security than that of surface web browsing to enable users to interact online without revealing their location or identity. TOR does so by encrypting the application layer of a communication protocol stack multiple times and then relays that data through several, normally at least three, randomly selected TOR relays [2]. At the end of the connected TOR network is the exit note where the encrypted traffic exits into the public destination. The exit node decrypts users' traffic so that it can reach their final destination on the regular Internet, which makes it the focal point for intelligence services to focus their efforts on intercepting user data. It is to be noted that the exit nodes are also able to be corrupted as anyone can set up an exit node, meaning malicious intent can occur at this stage using TOR. However, there are steps that can be taken to protect and avoid these "bad" exit nodes [32]. The darknet came to the attention of the general public and academia due to the rise of illicit trading on the darknet, the most infamous example being the online drugs market, "Silk Road". Online transactions have become a part of everyday life in society across the globe and are the foundations of almost all commerce. The first commerce deal facilitated by a computer network was in fact a marijuana drug deal between students at Stanford and the Massachusetts Institute of Technology (MIT) [27]. This is often refuted as the first-ever e-commerce purchase due to the deal was only facilitated through the Internet, but no financial transaction was conducted using the Internet. Smithsonian Magazine has labelled a sale of a "Sting" CD being the first-ever online purchase, with credit card information being entered online using data encryption.

The remainder of this chapter is structured as follow. Section 2 discusses the challenges that the darknet present to law enforcement and other security agencies. that cryptocurrencies present. Section 3 briefly discusses the some of the beneficial applications of the darknet. Section 4 explores some of the existing solutions currently deployed to address the challenges of the darknet. Finally, concluding remarks are provided in Sect. 5.

2 Challenges

2.1 Challenges of Cryptocurrencies and Cryptomarket

One of the leading driving points which were key to developing an international market on the darknet was the utilisation of cryptocurrencies. Standard legal tender in most countries is called "Fiat Money" with obvious examples being the dollar, Pound, and Euro. Fiat money is issued by a government and is heavily regulated by centralised banks and offers an alternative to commodity money. Commodity money is created through the possession and sale of precious metals such as gold and silver, whereas fiat money is based on the "creditworthiness" of the issuing government [5]. Cryptocurrencies, on the other hand, are a virtual currency that has been utilised as a medium of exchange on the Internet. Unlike fiat currencies, cryptocurrencies are not regulated or controlled by any central authority. As cryptocurrencies are not limited to a certain region, they are easily accessible all over the globe and offer instantaneous settlements and transfers [14]. The main aspect which has promoted the rise of cryptocurrencies across the Internet is the privacy element of their transactions. Using encryptions and other cryptographic techniques ensures that "participants and their activities remain hidden to the desired extent on the network" [35].

One of the prevalent sources of financially malicious aspects of the darknet is the selling of passwords and credit card accounts often sold for a fraction of the available balance on each account [31]. The bulk of the stolen sensitive information is taken from large-scale data breaches within businesses from around the globe. The largest of this was in 2013, when Adobe was the target of a large-scale data breach in which over 150 million user records were exposed. Each user had their names, identification, passwords, and debit and credit card information released across the dark web. The result was likely to provide a surge in sales of stolen data across the cryptomarkets as well as the legal fees and settlement claims that Adobe had to pay [37]. The Adobe hack damages to the company were relatively low, equating to just over \$2million. However, comparatively, there have been other large style data breaches that have ended up costing the companies a significant amount of money, affecting stock markets and even paving the way for cyber-espionage from foreign governments with political motivations. Another example was back in 2011 when the PlayStation Network was hacked. The hack caused a subsequent outage for all PlayStation services. The attack successfully breached Sony's security measures and gained access to all 77 million users' real names, postal addresses, country, email address, date of birth, username, passwords, and security answers [34]. This mass breach of personal information and the subsequent selling of this data across the cryptomarket would have undoubtedly made the victims vulnerable to a wide range of attacks, from email, telephone, and postal scams. The attack had cost Sony roughly \$171 million and at the time was one of the largest hacks [34]. Since then, however, in the US alone, there have been numerous additional large-scale attacks with the Target 2013 and Home Depot 2014 attacks being another two examples. Between the two attacks, over 96 million customers' names and credit card information were stolen equating to over $202 million and $179 million, respectively. With both of these cases, in the following investigation, it was detected that the breaches were directly linked to the Russian state and had political motivations as well as the obvious financial intentions. It came at a time of heightened tensions between the Western countries (US and Europe) against Russia following the aggression against Ukraine. The stolen batches of credit cards were uploaded to the dark web in groups named "American Sanctions" and "European Sanctions" [26].

2.2   Anonymity

The combination of TOR and cryptocurrencies such as Bitcoin provide the ability to sell illegal items such as narcotics, weapons, fake identification documents [20] among other services, anonymously and with relatively little risk of detection [1]. Although "Silk Road", the first dark net marketplace, was closed down by the FBI, the marketplace is reoccurring under new versions and names and remains targeted by intelligence authorities. There is significant research that suggests that the use of the darknet for buying illicit drugs on the crypto markets is rising exponentially, despite interventions by intelligence services and crypto market scams. However, the turnover from the crypto-drug market is still considerably less than that made from offline drug sales. Evidence suggests that cryptomarkets might be supplying the offline distribution as over 25% of the drug transactions on the cryptomarkets were of a value of over $1000. This suggests they have been bought for supplies or wholesale purposes [17]. The decentralization of these cryptomarkets, the international reach, and the inherent anonymity provide, flexibility and durability of this infrastructure, and with the encrypted exchange between the "facilitator" and "broker" being encrypted make the cryptomarkets an extraordinarily difficult target for law enforcement [25].

2.3   Hacking and Hacking Forums

One of the most potent tiers on the darknet is the hacking services. Although the term "hackers" is often disputed, hacking generally refers to the activities which seek to compromise digital devices ranging from computers and phones to entire networks. Although hacking is often motivated by malicious intent by what is known as a "black hat hacker" with the aim of financial gain and spying, it can also be done under ethical motivations by those referred to as a "white hat hacker", or a "grey hat hacker" who is a combination of the two [24]. Hacking for financial gain is the most prevalent form of motivation with a wide variety of systems being utilised and targeted. Porolli [33] states that there are options for low-skilled users of the darknet to utilise the hacking community for malicious intentions [33]. Due to the proliferation of hackers across the darknet, within the cryptomarkets, individuals can purchase a variety of hacking services. One of the most prevalent of these is ransomware packages. Ransomware is a form of malicious software (malware) that encrypts the target's files. The attacker then demands a ransom to restore the victim's access to data.

The ransom usually ranges from a few hundred pounds to thousands depending on the target and is payable to the cybercriminal via Bitcoin or other cryptocurrencies. One of the most common methods for disseminating ransomware is using phishing-style scams, which are often attached to carefully crafted emails that attempt to masquerade the ransomware file for one they would trust [13]. This style of attacks requires little to no experience to utilise and can be purchased from cryptomarkets with a singular payment or in monthly and annual instalments. Although there are ethical hackers or "white hats" who actively use the hacking forums for research and relatable news, these hacking forums are mainly used for illicit intentions. One of the largest hacking forums is "FreeHacks", a Russian-based community that actively aims to gather resources to maximise efficiency and knowledge dissemination [15]. The forum has efficiently divided up into different categories with links to various related sub-forums. These categories are largely focused on illegal activities such as hacking, "carding", Botnet, Brutus, Malware, Financial operations, False Documentation, and even a community judicial system or Blacklist [15]. Although the darknet can be used as a platform for cyber security education and as an international community, the forums are largely driven by illegal activities.

2.4 Terrorism

Terrorists have been active on the surface web since the 1990s. However, in a similar way to organised criminal groups, terror groups have found that the surface web is often too risky with a greater chance of being monitored, tracked, and found. Quilliam Foundation state, "The terrorist material reappears on the Internet as quickly as it is banished and this policy risks driving fanatics onto the 'dark web' where they are significantly harder to track. Moreover, Islamist forums and chat rooms are still widely available on the surface web, however, a large proportion of more extremist Islamic discourse now takes place on the dark web" [18]. The migration of much of the communication between ISIS members was accelerated following the 2015 attacks in Paris. As a retaliation to the attack which killed 130 people across Paris, the international hacktivist group/collective called Anonymous launched a "digital war" against ISIS.

OpParis as it came to be known, was Anonymous' largest combined effort to date where they also urged other hackers from across the globe to assist and join in with the attack on Jihadism online [22]. Anonymous used a variety of tools to hinder ISIS' online presence by cracking passwords and resetting their emails which in turn would seize control of the account. If this were not possible, they would flag these accounts to Twitter or Facebook so they can be closed down. Additionally, they used a wide variety of "cracking" tools and attacks such as distributed denial of service (DDOS) attacks, which overwhelm a website with large amounts of traffic. This in turn causes the website to crash. A few months after the attack started, a partial victory was declared by Anonymous, which had successfully brought down roughly 150 websites, flagged over one hundred thousand Twitter accounts, and reported a further 5000 propaganda videos [16].

Terror groups use the dark web for more of the same purposes as they used the surface web just with the added privacy and anonymity that the TOR encryption provides. These activities include propaganda dissemination, recruitment, coordination, and training. Just two days after the Paris attacks, ISIS posted a message containing details and instructions on moving activity to the darknet. Shortly after, al-Qaeda disseminated a manual entitled "Tor Browser Security Guidelines", which offered instructions on installation and using the TOR browser and means to avoid identification by intelligence agencies [39]. Additionally, the use of the darknet by terrorists is of great concern due to the vast material and opportunities that the darknet supplies. This includes the cryptomarkets discussed above. It is well known that the darknet cryptomarkets or "black markets" can contain a wide variety of items, from drugs, guns, and hitmen for hire. However, in 2016 during a briefing held by the then President, Barack Obama, it was described how a terrorist group had bought isotopes through brokers on the dark web [19]. This worst-case scenario is abetted by the recent usage of chemical weapons by non-state actors and ISIS in Syria with the use of mustard gas against civilians and "rebels". The darknet crypto markets and the anonymity provided by the TOR browser and cryptocurrencies could provide a greater risk of terror groups gaining access to chemical, biological or nuclear material needed to manufacture a "dirty" bomb or weapon of mass destruction in the future [3].

2.5 Challenges of Policing Cybercrime

The first step when it comes to policing cyber-crime is attempting to identify the suspect, ideally through tracing their Internet Protocol (IP) address. However, although accessing the TOR browser may be visible by law enforcement and an individual's internet service provider, most users will access the TOR browser using a Virtual Private Network (VPN) to avoid raising suspicion. A VPN creates an encrypted tunnel for the data that the user sends and receives and is only visible by the VPN provider. Combining the use of a VPN with that of the TOR browser which uses encrypted traffic nodes ensures a high degree of privacy and anonymity in terms of the user's location and what content they have accessed [38]. Details of exact tools and methods currently being utilised by police agencies are for obvious reason largely not publicly available. However, the main known method utilised by police in their cybercrime investigations is what is known as Open-Source Intelligence or OSINT [30]. OSINT requires investigators to search for "breadcrumbs" of information that may lead to the real identity. These are often left unintentionally through human error. This can come from content distribution, contact by the user, and social networking to name a few. This process was what secured the arrest and prosecution of the creator of the Silk Road, Ross Ulbricht, whose personally identifiable email address was used to advertise the marketplace. However, the "pop-up" nature of the cryptomarket places where they have a fast rate in which they open and close, due to servers being hacked, raided, or abandoned, or set up as an "exit scam" impede the efforts of law enforcement in identifying the users.

3   Positive Application

Although the research and news relating to the darknet are predominantly focused on illicit activities, there are some positive aspects of the darknet. As previously discussed, the darknet is a very sociable environment with community pages and forums being an obvious example. The need for discourse whilst remaining anonymous has also led journalists and whistle-blowers to share their social and political beliefs openly without fear of retribution [28]. This sharing of information has become especially common in countries with strong state censorship and surveillance against political activists, freedom fighters, and journalists [21]. This was witnessed during the Egyptian riots when journalists and activists were able to bypass the government censorship by using the TOR browser and is now even recommended as part of the "survival kit" for journalists in the industry. There has also been a backlash towards the mass collection of users' personal data and the selling of such information to third parties whether these are politically motivated or for financial gains. This especially came to light during the Cambridge Analytica scandal, in which it was revealed that psychological profiles were generated by acquiring Facebook's "private" data of tens of millions of users. This data was then used to tailor advertisements and propaganda by certain parties in both the Trump/Hilary election in the United States, and the Brexit campaign in the UK [6].

4   Examples of the Existing Solutions

4.1   Undercover Policing

Undercover policing has been used for some time in the attempt to counter and prosecute cybercrimes. However, the ability to infiltrate online forums to gain enough substantial and credible evidence that is required for a successful prosecution is notoriously difficult. Many of these operations require the police investigators to establish a surveillance operation in which the officers covertly act as administrators or moderators of these illegal forums [23]. This method has been very successful in tackling multiple illicit activities on the darknet, including catching paedophiles by posing as children within the communities. Arguably, one of the most successful undercover cybercrime operations was when a United States-based law enforcement agency took over an account of a staff member of the silk road. Whilst maintaining their cover during the closure of the Silk Road, they were then invited to contribute to the creation of the second "Silk Road", Hansa. During the operation, network monitoring equipment was installed which eventually led the Netherland National High-Tech Crime Unit (NHTCU) to identify the TOR-protected server that ran Hansa. The NHTCU then continued monitoring the traffic through the server for a month with officers posing as administrators of the site, during which the FBI located another server that was responsible for the Alphabay marketplace, one of the largest in the world [8]. The coordination between the NHTCU and the FBI resulted in over 27,000 transactions being analysed, and data was obtained from 420,000 users including 10,000 home addresses [7].

4.2 Machine Learning

Machine Leaning (ML) is a technology that is increasingly being researched and developed. Typically, law enforcement need to link identities on the darknet to the surface web to uncover an individual's true identity. The current issue with this method is the sheer volume of data that needs to be sifted through. The automation of the linking process allows the introduction of ML algorithms to scour the darknet and compute for the similarity between users across a variety of different forums. The algorithm uses at least three separate steps during its linking process. The first step is to analyse data from at least two subgroups and search for similarities in profile information, including username similarities and spelling. The next step would be to identify content similarity across the data sets and similarities in the users' circle of people that they interact with and the topics which are discussed. These steps are then used to provide probability scores of linking the digital personas to that of a real-world or surface web profile [12]. The main program that is currently being utilised is the Defense Advanced Research Projects Agency's (DARPA) Memex program. This program is currently being used by at least 30 agencies worldwide [29]. The Memex tool has had great success amongst one of its largest users, a Human Trafficking Response Unit (HTRU) in the Manhattan District Attorney's Office. In 2017 alone, it was reported that over 6000 arrests were made for human trafficking and that Memex was used in 271 human trafficking investigations. Furthermore, it increased prostitution-related arrests from 15 to 300 per year [12].

5 Concluding Remarks

The anonymity and privacy elements are one of the largest draws towards the TOR browser, as general surface web browsing is trackable and recorded, with tailored advertisements using cookies and constructed targeting based on demographics, recent purchases, and history [36]. Although the darknet was created for military and intelligence operators as a form of data sharing and communication, the privacy and anonymity that the browser provides have been utilised and exploited by the criminal world. Whilst governments spend considerable amounts of money and resources policing the darknet, opening it up to the public domain was intentional as a means of further "masking" or hiding their military and intelligence operatives in the darknet. Similarly, while law enforcement agencies shut down the original Silk Road marketplace, newer and decentralised marketplaces quickly reappear and are often better hidden than the previous one. This essentially creates a virtual game of cat-and-mouse amongst the law enforcement and the criminal community, where the law enforcement agencies are typically only able to act in a reactive manner rather than preventative.

The same can be said for the large scale of financially malicious attacks which are created, bought, sold, and disseminated throughout the darknet, costing the global economy over a trillion dollars, and this figure is rising exponentially in recent years [11]. It is important that the darknet is available for the use of journalists and activists, especially in regions of censorship and imprisonment threats, and should act as a lawful freedom of expression. However, the obvious use of the darknet for illicit materials calls for an international solution to combat the nefarious and illicit activities undertaken on the darknet, with clear guidance for international cooperation to aid in the successful prosecution of these individuals. This solution may stem from advanced ML techniques or artificial intelligence tools such as Memex. It is to be noted, however, that the criminal world will likely utilise and develop its own ML techniques to develop more complex means of cracking tools which will undoubtedly lead to larger-scale attacks in the near future.

References

1. Barrat MJ, Ferris AJ, Winstock AR (2014) Use of silk road, the online drug marketplace, in the UK, Australia, and the USA. Addiction 109:774–783
2. Bee (2020) How does tor really work? the definitive visual guide. Skerritt.blog. https://skerritt. blog/how-does-tor-really-work/. Accessed 10 May 2021
3. Besheer M (2017) UN: terrorists using 'Dark Web' in pursuit of WMDs. Voice Am. https://www.voanews.com/europe/un-terrorists-using-dark-web-pursuit-wmds. Accessed 10 May 2021
4. Biddle P, England P, Peinado M, Willman B (2002) The darknet and the future of content distribution. In: ACM Workshop on digital rights management, vol 6, p 54
5. CFL (2015) Fiat money. Corporate Finance Institute. https://corporatefinanceinstitute.com/res ources/knowledge/economics/fiat-money-currency/. Accessed 10 May 2021
6. Confessore N (2018) Cambridge analytica and facebook: the scandal and the fallout so far. The New York Times 4:2018
7. Davies G (2020) Shining a light on policing of the dark web: an analysis of UK investigatory powers. J Crim Law 84(5):407–426
8. The United States Department of Justice (2017) AlphaBay, the largest online 'Dark Market,' shut down. https://www.justice.gov/opa/pr/alphabay-largest-online-dark-market-shut-down. Accessed 10 May 2021
9. Deyan G (2021) How much of the internet is the dark web in 2021?. TechJury. https://techjury. net/blog/how-much-of-the-internet-is-the-dark-web/#gref. Accessed 10 May 2021

10. Dingledine R, Mathewson N, Syverson P (2004) Tor: The second-generation onion router. Naval Research Lab Washington DC.
11. Fadilpasic S (2020) Cybercrime is now a trillion-dollar cost to the global economy. https://bit. ly/3jkBRo2. Accessed 10 May 2021
12. Foy K (2019) Artificial intelligence shines light on the dark web. Massachusetts Institute of Technology. https://news.mit.edu/2019/lincoln-laboratory-artificial-intelligence-hel ping-investigators-fight-dark-web-crime-0513. Accessed 10 May 2021
13. Fruhlinger J (2020) Ransomware explained: how it works and how to remove it. CSO online. https://www.csoonline.com/article/3236183/what-is-ransomware-how-it-works-and-how-to-remove-it.html. Accessed 10 May 2021
14. Goyal S (2017) The difference between fiat money and cryptocurrencies. https://www. fxempire.com/education/article/the-difference-between-fiat-money-and-cryptocurrencies-520616. Accessed 10 May 2021
15. Gurran D (2018) My deep dive into one of the largest dark web hacking forums | Dylan Curran. The guardian. https://www.theguardian.com/commentisfree/2018/jul/24/dar knet-dark-web-hacking-forum-internet-safety. Accessed 10 May 2021
16. Hern A (2015) Anonymous 'at war' with Isis, hacktivist group confirms. The guardian. https://www.theguardian.com/technology/2015/nov/17/anonymous-war-isis-hackti vist-group-confirms. Accessed 10 May 2021
17. Hoorens S (2018) Online drugs trade growing but still dwarfed by traditional markets. https:// www.rand.org/randeurope/research/projects/online-drugs-trade-trafficking.html. Accessed 10 May 2021
18. Hussain G, Saltman EM (2014) Jihad trending: a comprehensive analysis of online extremism and how to counter it. Quilliam
19. Hutton R (2016) Bloomberg—Are you a robot?. Bloomberg. https://www.bloomberg.com/ news/articles/2016-04-01/nuclear-drones-from-dark-web-cited-by-obama-in-terror-scenario. Accessed 10 May 2021
20. Jacobs T (2021) Inside silk road—The dark web's first underground drugs network. The sun. https://www.thesun.co.uk/news/14464144/ebay-vice-illegal-dark-web-drugs/. Accessed 10 May 2021
21. Jardine E (2015) The dark web Dilemma: Tor, anonymity and online policing. SSRN. https:// papers.ssrn.com/sol3/papers.cfm?abstract_id=2667711. Accessed: 10 May 2021
22. Kittle B (2015) Hacker group Anonymous declares digital war on ISIS after Paris attacks. https://www.prweek.com/article/1372998/hacker-group-anonymous-declares-digital-war-isis-paris-attacks. Accessed 10 May 2021
23. Lusthaus J (2013) How organised is organised cybercrime? Global Crime 14(1):52–60
24. Malwarebytes (2020) Hacking definition: what is hacking?. https://www.malwarebytes.com/ hacker/. Accessed 10 May 2021
25. Martin J (2014) Drugs on the dark net. Springer
26. McCartney J (2015) The home depot hack: how, why and what we can learn. TechRadar. https://www.techradar.com/uk/news/internet/the-home-depot-hack-how-why-and-what-we-can-learn-1268333. Accessed 10 May 2021
27. Misulonas J (2021) Researchers: the first thing ever sold on the internet was Marijuana. Civilized. https://bit.ly/3h3moaG. Accessed 10 May 2021
28. Moore D, Rid T (2016) Cryptopolitik and the darknet. Survival 58(1):7–38
29. Odom J (2014) Memex (Archived). DARPA. https://www.darpa.mil/program/memex#:~:text= Memex%20seeks%20to%20develop%20software,relevant%20to%20their%20individual% 20interests. Accessed 10 May 2021
30. OSINT Academy (2020) Dark web searching. OSINT Combine. https://www.osintcombine. com/post/dark-web-searching. Accessed 10 May 2021
31. Owaida A (2020) How much is your personal data worth on the dark web? https://www.wel ivesecurity.com/2020/08/03/how-much-is-your-personal-data-worth-dark-web/. Accessed 10 May 2021

32. Phillips G (2019) 6 ways to stay safe from compromised tor exit nodes. https://www.mak euseof.com/tag/priority-wretched-hive-scum-villainy-5-ways-stay-safe-bad-tor-exit-nodes/. Accessed 10 May 2021
33. Porolli M (2019) Cybercrime black markets: dark web services and their prices. https://www.welivesecurity.com/2019/01/31/cybercrime-black-markets-dark-web-services-and-pri ces/. Accessed 10 May 2021
34. Saqibjazz (2020) The 5 biggest and greatest hacks of all time. Medium. https://medium.com/@saqibjazz28/the-5-biggest-and-greatest-hacks-of-all-time-6e1cafa8e10f. Accessed 10 May 2021
35. Seth S (2020) Explaining the crypto in cryptocurrency. Investopedia. https://www.investope dia.com/tech/explaining-crypto-cryptocurrency/. Accessed 10 May 2021
36. Smith L (2020) 8 audience targeting strategies from digital marketing experts. https://www.wordstream.com/blog/ws/2019/04/15/audience-targeting. Accessed 10 May 2021
37. Swinhoe D (2021) The 15 biggest data breaches of the 21st century. https://www.csoonline.com/article/2130877/the-biggest-data-breaches-of-the-21st-century.html. Accessed 10 May 2021
38. Symanovich S (2021) What is a VPN?. https://us.norton.com/internetsecurity-privacy-what-is-a-vpn.html. Accessed 10 May 2021
39. Weimann G (2016) Terrorist migration to the dark web. Perspect Terror 10(3):40–44

Detection and Binary Classification of Spear-Phishing Emails in Organizations Using a Hybrid Machine Learning Approach

Popoola Favourite Akinwale and Hamid Jahankhani

Abstract In recent years, e-mails have become the most commonly used information and communication technology in organization. The email technology is susceptible to phishing attacks. Phishing attacks are a trending cybercrime activity and have caused a lot of financial loss using social engineered communication that are transmitted to individuals by cybercriminals aimed at tricking users to share sensitive information, with emails being the prevalent attack vector. Spear-phishing is a deceptive social engineering tool used to capture targeted victim's personal data or information. Social engineering is a cyber threat that aims to use psychological techniques to manipulate or trick people to expose their personal/sensitive information. In this research paper, five supervised classification algorithms (Logistic Regression, Random Forest, Decision Tree, Support Vector Machine and K-Nearest Neighbour) are discussed and compared. A pre-processing system has been designed to collect and extract the features and patterns of spear-phishing emails from the email header and email body. In addition, using the top two classifiers, Logistic Regression and Decision Tree, a novel technique for detecting spear-phishing emails by utilizing multiple classifiers, and the findings showed the hybrid classifier provides a more precise results for the features of detecting spear-phishing emails with a 99.8% accuracy with a low-false positive and false negative rate.

Keywords Phishing · Spear-phishing · Social engineering · Machine learning · Email · Classification · Hybrid classifier

P. F. Akinwale · H. Jahankhani (✉)
Northumbria University, London, UK
e-mail: Hamid.jahankhani@northumbria.ac.uk

215

1 Introduction

1.1 Introduction

Increased digitization of the various areas of industry and human practices had led to a spike in cybercrime worldwide. The current threats in the cyber world are a growing security issue, as cybercriminals have increasingly found innovative ways to take advantage of specific targets. Owing to the growth of the internet architecture and online commerce, online users are required to input personal identifiable information on some websites. This information is targeted by cybercriminals, and a new form of attack began to emerge in the mid-1990s called Phishing. For years, the threat to the cyber world, Phishing, has been seen as a big challenge rising exponentially every year. It is considered an illegal act that combines social engineering with technological techniques for stealing sensitive information [1].

Social Engineering is a cyber threat targeted at exploiting or deceiving unsuspecting users into revealing their sensitive information through psychological means. Social engineering has evolved over the years and has been a primary attack vector employed by cybercriminals to carry out a range of targeted attacks on individuals in organizations of interest.

The Anti-Phishing Work Group (APWG), an international body was created in 2013, to help in raising awareness about the online threats ravaging the cyberspace [2]. This body regularly disseminates reports to the online community, this report contains up to date information on the current level of cybercriminals, recent cyber-attacks and the current approaches of the cybercriminals. The latest report stating that in the second quarter of 2020, over 146,994 phishing websites were discovered, 289,012 phishing attempts were reported, with the biggest category of phishing being Webmails & Software-as-a-Service (SaaS) and 78% of discovered phishing websites using SSl protection [3].

In 2017, SiteLock mentioned that about 1.86 billion webpages are available on the internet, about 18.5 million web pages (1% of available web pages) are infected with malware by cyber criminals weekly and the average website is attacked 44 times daily [4].

A recent report from google states that over the past 4 years there has been a rapid increase in phishing websites and as at May 2020, google had over 2 million phishing websites in its database that were deemed dangerous [5].

Presently there exists a strong email dependency in how individuals and organizations communicate. Which has become a double-edged sword for the same infrastructure that facilitates efficient communication for organizations and individuals also offers an incentive for malicious actors to participate in cyber-crime related activities.

There has been a rapid advancement & development in the Information Technology industry in recent years, this has led to the enormously number of users who perform online transactions and the surge of financial value exchange performed through electronic means i.e. Private Payment Gateway. Although, the convenience this has brought for both internet users and organisations regarding online transactions, has also come with a price. The online threat known as spear-phishing materializes.

The increasing frequency of spear-phishing attacks in the organizational context and electronic commerce industry has unfortunately led to legitimate users losing trust in online commerce transactions and organisations losing loads of money to cybercrimes.

It is often difficult to track and detect a phishing attack, most users do not realise that they have given out sensitive data, personal identifiable information to cybercriminals, as the vast majority of phishing attacks go unnoticed.

The emphasis in this paper is to create an intelligent email classifier model that is capable of detecting spear-phishing emails at an early stage in the phishing process. A lot of research has shown that the human factor is not a favoured option to fighting phishing attacks, the model proposed by this paper will eliminate the spear-phishing emails before it is received by humans.

This approach to spear-phishing detection employs the knowledge discovery in database methodology and machine learning classification techniques to create an intelligent model from the Enron email dataset containing legitimate and spear-phishing emails. The model isolates and minimizes the essential features, the features are extracted from the email header and email body, after which the features and contents of the email go through a series of pre-processing processes i.e. tokenization, weighted word frequency and linguistic analysis. After completion, they are implemented on five classification algorithms to determine the best classifiers. The top two classifiers from the experiment are merged together to create one hybrid classification model to help in the detection of spear-phishing emails.

2   Literature Review

2.1   Introduction

The work in this chapter is focused on detecting spear-phishing emails with machine learning techniques. Machine learning has increasingly become popular based on its accuracy when it is utilized and is levels ahead of similar detection approaches available (i.e. List based, Heuristics, Visual Similarity, Authentication protocols, Email-Spam Filtering) in detection of phishing emails.

This chapter offers a comprehensive explanation and investigation of what a phishing attack is, the life cycle of phishing attacks and a detailed investigation of existing- anti-phishing techniques. It also presents a number of algorithms available for the detection and prediction of spear-phishing emails, critically analyses the advantages and disadvantages and provides justification for selecting a hybrid machine learning model for this thesis.

2.2 Phishing

The word phishing originates from Fishing, fishing as defined by Merriam webster is the sport or business of catching fish. The phishing process starts when the phisher (Fisherman) throws out a bait in a bid to fish for data, the techniques is completed when unsuspecting internet users (victims) bites the hook released by the phisher, the data of the unsuspecting internet user (victim) is gotten by the phisher once that happens, who may use it to perform illegal activities.

It is a cybercrime where unsuspecting internet users (victims) are misled into giving out sensitive and personal identifiable information such as login credentials and financial account details to cybercriminals who have malicious intentions.

Cybercriminals employ social engineering and technical skills to mislead unsuspecting internet users (victims) that assume the internet communication is with a legitimate entity.

Phishing as defined by APWG is a crime where criminals employ both social engineering and technical subterfuge to steal consumers' financial account credentials and personal identifiable data [3].

Cisco defines phishing attacks as the practise of sending fraudulent communication that appears to come from a trustworthy source with the purpose of compromising all available data sources or install a malware to compromise the victims machine and connected systems [6].

2.2.1 Types of Phishing Attacks

1. Email Phishing: This is the most common type of phishing attack; cybercriminals attempt to obtain sensitive and personal data by registering an illegitimate domain that imitates a legitimate organization and send out thousands of generic emails to unsuspecting internet users.
2. Spear-phishing: This is a type of email phishing; it is more directed and targets a specific person(s) rather than a wide range of people. The word spear connotes that it focuses on a particular user or unit in an organisation. Cybercriminals tend to do a bit of reconnaissance on the targeted individual before starting the spear-phishing attack. This technique employed by the cybercriminals is used to penetrate defences put in place by organizations, in a bid to carry out targeted attacks. A recent survey carried out by SANS Institute identified the success rate of phishing attacks on enterprise networks to be 95% [6]. This thesis paper will be proposing solutions to help curb this type of attack.
3. Whaling: This is another type of email phishing and is considered to be specific and focused as it targets senior and high-profile business executives. It is similar in nature to spear-phishing; it involves targeting a particular individual and formulating communications that appeal to the targeted individual. This attack is a lot more subtle as the cybercriminals spend a lot of time profiling the targeted individual before initiating the attack. Whaling attack is considered as high risk

due to the access senior and high-level executives have to sensitive organization information [7].

4. Smishing: This is an act where cybercriminals try to lure users into visiting malicious websites, downloading malware, giving out sensitive and private information through text messages [8]. The text messages are made to look like they are from a legitimate and credible source i.e. financial institution, service providers. The smishing messages are used by the cybercriminals to incite fear or interest [9].

5. Vishing: Vishing is a blend of two words Voice and Phishing. Cybercriminals use an internet phone popularly known as voice over IP (VOIP), instead of the regular illegitimate email, illegitimate website or phone call and text message to deceive internet users into giving out their valuable and sensitive information. The attacker's spoofs a legitimate phone number making the victim assume it's from a legitimate source and uses the mixture of scare tactics with emotional manipulation to deceive the victims into giving out sensitive and valuable information [10].

6. Baiting: This is a fairly old technique used by cybercriminals. This technique involves the criminal placing an infectious device i.e. flash drive, hard drive, floppy disk, cd in a location that is easily accessible to potential victims. This technique relies on the curiosity of the victim wanting to find out the content of the infectious device he/she found lying around without the prior knowledge that the said device was placed there by cybercriminals. Once the victim connects the infectious devices into a system. The malicious content on the device automatically installs itself on the victim's computer, thereby giving the cybercriminal access to the victim's systems and organizations internal network.

7. Wi-Phishing: This phishing technique targets unsuspecting users of wireless network devices i.e. Smartphones, Laptops. Criminals employ this technique in populated hot places such as hotels, coffee shops, restaurants, cybercafes and the underground, as internet users tend to search for available wireless networks in these areas. Unsuspecting internet users connect to these compromised wireless networks and their information is gotten by the cyber criminals.

8. Pharming: Pharming is a technique used to gain unsuspecting internet users sensitive and valuable information. It is carried out when a user visits a spoofed website i.e. A user redirected to a hoax website that seems like a legitimate one. The unsuspecting internet user then enters his/her details from a prompt asking for the victim's details, hereby giving the attacker his/her details.

9. Angler Phishing: This is a new approach that cybercriminals have adopted. Since the introduction of social media, loads of internet users have a social media account where they disclose seemingly harmless information. Cybercriminals now monitor the social media accounts of their victims, create fake accounts on social media and persuade social media users into divulging their sensitive and valuable information [11].

2.2.2 Phishing Background

The threat phishing makes use of spoofed emails and illegitimate websites to trick unsuspecting victims into voluntarily giving out their sensitive and valuable information. Phishing is coined from the word Fishing and was first introduced in the United States of America, where phishers started, figuratively to fish illicitly for victims personal and sensitive data. The phisher throws a bait in the form of illegitimate websites, emails and waits for his/her victim to take a bite, thus capturing the victims information (fish) [12].

The origin of phishing can be traced back to the early 1970s, where hackers traded the word f for ph in an earlier form of crime named "Phone Phreaking", hacker were hacking telephone systems to make illegal and free phone calls [13].

John Draper, popularly known as Captain Crush was the first hacker recorded to have carried out Phone Phreaking, he made use of a toy whistle wrapped up in a Cap'n Cruch breakfast cereal box (famously known as the Blue Box) emitting a frequency of 2600 Hz which allowed him to make free phone calls on the American AT&T lines from any telephone point in the United states of America [14].

In 1996, phishing became an internationally recognised cybercrime due to a successful phishing campaign carried out on an American internet service provider called America Online (AOL). Often regarded as the first phishing attack [15], the hackers generated random credit card details which were used in creating AOL accounts. After creating AOL accounts, the hackers made use of AOL's instant messages in AOL chat rooms & emails to reach unsuspecting AOL users. Acting as AOL employees the hackers asked users to update billing information, verify certain customer information such as personal identifiable data and financial account details. AOL sent out emergency emails to their customers and instant messaging warnings in a bid to inform and protect users from providing sensitive information to the attackers [16].

2.2.3 Lifecycle of Phishing Attack

Jain and Gupta [17] explained the steps involved in a successful phishing campaign. There are several ways in which a phishing campaign can be initiated, the most popular is via email and others such as social media websites, text messages, instant messaging and online blogs are viable options.

A successful phishing attack contains the following:

Step 1: The website of a legitimate organization is cloned to create an illegitimate website with visual similarities to that of the legitimate website in a bid to fool unsuspecting users.

Step 2: The link of the illegitimate website is placed in an email composed by the attacker and sent out to a number of potential victims. A targeted attack will be directed to a specific individual or department in an organization.

Step 3: The potential victim receives the email and clicks on the link placed by the attacker. The victim is directed to the malicious/illegitimate websites and inputs his/her sensitive or valuable information.

Step 4: The victims information is received by the attacker, or a malicious software runs and installs itself on the compromised victims system. The attacker makes use of the victims information to commit cybercrimes or gains access to the organization enterprise network in the case of a spear-phishing/whaling attack.

2.3   Existing Anti-Phishing and Detection Techniques

Over the years, approaches taken by organisations and individuals to mitigate phishing attacks have proved abortive, as serious breach of user confidentiality, organizational integrity and service availability have seen a rise in occurrence.

Researchers and scientists have proposed many techniques to help combat phishing attacks and protect individuals as well as organizations. These techniques are categorized as non-technical and technical approaches, and can both be employed by individuals and organizations.

The non-technical approach primarily involves legal, simulations, awareness and training programs which are directed at internet users to explain and inform them about current phishing trends and procedures to take in handling a phishing attack.

However, the technical approach provides solutions that covers the scope of this paper, which is detecting phishing and spear-phishing attacks and does not require user intervention.

2.3.1   Non-technical Anti-phishing Approach

There are a variety of non-technical approaches available to help combat and mitigate phishing attacks.

Legal Legislations

Having stringent legal legislation and legal recourse on phishing attacks is essential in deterring cyber criminals from performing phishing attacks. World governments failed to realise this earlier on, thus there were no laws opposing phishing.

In 2005, the first ever instance of an Anti-Phishing legislation was issued in the state of California in the United States of America [18]. The anti-phishing legislation stated that it is against the law to use any electronic means such as emails, website or alternative methods to request or canvass for information from online users by claiming oneself as a business without the authority of that business [19].

A few states in the United States of America such as Texas followed suit in creating anti-phishing legislation. Thus, providing organizations with the legal defense to fight cyber criminals (phishers). The federal government as a whole and other states who didn't have an anti-phishing legislation could only prosecute perpetrators under the general available computing laws.

In light of this new development in the United States of America, the United Kingdom emulated the approach by enacting more stringent cyber crime penalties along with identity theft and fraud in 2006 called the fraud act [19].

In accordance with the fraud act, an increased prison sentence of up to 10 years was given to cybercriminals for cyber crimes. The fraud act also contains laws prohibiting the possession of phishing websites with the goal of deceiving individuals and committing fraud. By 2010, the government of Canada also introduced an anti-spam law to shield canadians against cyber crimes which came into force on july 1, 2014 [20].

Although these laws are available now and serve as deterrent against phishing attacks, they pose little or no threats to cyber criminals with advanced persistent attacks. As such, it is important to establish and incorporate other phishing prevention strategies.

Education, Training and Awareness

A significant approach in mitigating phishing is user education, this is often regarded as the first line in defending against phishing attacks. Organisations primarily implement this approach to have training set up for their employees.

As organisation employees are customers that are targeted by phishers to effect cyber crimes. A successful phishing attack on an organisation can result in the financial loss, confidential personal identifiable information being stolen, loss of intellectual property which may lead to the loss of millions in monetary value. Thereby, reiterating the importance of employee awareness program as part of organizational training [16].

A number of different approaches ranging from games to simulated phishing emails can provide this form of training. The simulated phishing emails have evolved into embedded methods of training.

Embedded training is a method of training that offers training information to users via a simulated phishing email. The goal of this approach is to identify the instance the user fell victim to a phishing email attack and provide a learning opportunity.

Strategically placed in the simulated phishing email is an embedded URL, when the user clicks the URL in the mail, the user is presented with the training content. The training content varies from a basic warning to a fully-edged teaching material on phishing. The deployment of the simulated phishing email is used as an avenue to provide training materials and can also be used to test the capacity of the user to detect phishing email [21].

The use of interactive games as a tool to teach users about phishing has been discussed among many scientists and researchers. Anti-phishing Phil, a game developed by [22], teaches users how to identify phishing urls in the game, the user takes on the character phil (a young fish) who wants to eat worms. Real worms were marked as legitimate urls while fake worms were marked as phishing urls.

A survey carried out by [23], had 20 participants presented with a list of 10 suspicious websites and were tasked to identify which one of them were legitimate urls before playing a prototype game similar to [22] anti-phish phil game. After completing the game, a fresh set of 10 suspicious urls were given to the participants to evaluate their legitimacy. The result of the survey proved that the participants performed better after playing the game.

Overall, several works have shown positive findings that interactive games may be a successful way to help users spot phishing urls, the research works also suggests that users who become aware of phishing urls through interactive games were notably better at detecting phishing urls than those who had been training through the reading the traditional security materials.

Phishing awareness campaigns are necessary but not sufficient enough to manage the issues of phishing fraud. Firstly, the method of approach employed by phisheers is constantly updated, thereby, existing training curriculums are rendered obsolete due to the latest techniques and approaches employed by the phishers. Moreover, for both the individuals and organization, the expense of such awareness programs is substantial and they need to be replicated and revised regularly. Furthermore, due to the relatively short life span of phishing urls the improvements in the sophistication of attacks will be miles ahead of any training awareness program, leaving organizations and individuals with lots of catching up to do. Finally, it is not possible to effectively deliver and provide adequate training to a large number of people, such as non-employees.

For these reasons, an intelligent system that provides online internet users the security required without the use of human interaction should be best practise in mitigating phishing attacks, the system will rapidly adjust to respond to changes in phishing attack vectors.

2.3.2 Technical Anti-Phishing Approach

There exist a variety of technical anti-phishing approaches available to detect and predict, and also counter phishing attacks. These approaches are based on manual or automated review of email contents without user intervention being required. The approaches often provide trade offs between coverage and accuracy and are seldomly designed to complement each other.

The approaches for the detection of phishing emails are divided into three major categories:

a.   Blacklist/Whitelist Detection
b.   Heuristics based Detection

c.   Machine learning based Detection.

Blacklist/Whitelist Detection

This approach for phishing detection employs using a database of both legitimate(whitelist) and phishing(blacklist) websites to prevent phishing attacks. The databases are often stored and updated on the user system or a local server.

- *Blacklist*

    This is the most primitive approach employed to tackle phishing attacks. In the past, organizations such as PhishTank made use of this approach by employing humans to manually verify the legitimacy of a website, if a website was deemed illegitimate, it would be added to the list of phishing url's (blacklist).

    The list of phishing urls was used by a series of protection softwares to alert users if they were visiting a potential malicious url. The manual verification process by humans incurred a significant amount of delay between the time where the url was first identified and applied to the blacklist.

    In comparison, there is a very low positive rate received on the blacklist approach, which prompts users to favor it relative to other provided approaches.

    Sheng et al. [24] carried out an assessment to measure the effectiveness of the blacklist approach in detecting phishing attacks, the process entailed making use of a series of fresh phishes i.e. a phishing attack launched in the last 30 min. The findings from the assessment showed that solutions that employed the blacklist approach are unsuccessful in stopping zero-day attacks, over 80% of the phishing attacks were not detected [24].

    The biggest drawback of this approach is the relatively short lifespan of phishing urls, the blacklist database would not include all phishing urls at a certain time. It is very possible for a phisher to have fooled multiple users between the start of the phishing attack and it's moment of discovery and also created a new phishing url [25].

- *Whitelist*

This approach aimed at building a registry of websites that have been checked and verified to belong to legitimate entities, as opposed to the blacklist approach. Owing to the large number of available websites hosted on the internet, establishing a worldwide whitelist registry for all legitimate websites was impossible.

In controlling zero day phishing attacks, the whitelist approach has proven successful as it delivers zero false positive. The key downside to the whitelist approach is that all urls that the user will visit in the future will be difficult to handle. Essentially, the system would assume a url to be an illegitimate url when a user attempts to visit a legitimate website that is not included in the whitelist registry, thereby raising the false negative score.

Chen and Guo [26] suggestested one of the early created whitelists, which was focused on the surfing of legitimate urls by users. The program tracks the login attempts of the user and if a recurrent login has been successfully performed, it requires the user to add the url into the whitelist. One apparent drawback of this program is that it suggests users are constantly communicating with trustworthy urls which sadly is not always the case.

Wang et al. [27] suggested APWL, a user crested Anti Phishing Whitelist. Pattern matching was employed by APWL to detect sensitive personal identifiable information such as social security number and blocked the user from submitting such data to a non-whitelisted url. In addition, except the user manually inputs the url's to the list of allowed url's (whitelist), APWL did not allow the user to circumvent APWL blocking. Performance analysis revealed that only a slight performance overhead was generated by the APWL program.

The listed whitelist approaches focus largely on users identifying urls which they trust and blocking users from sharing sensitive user information from potentially unsafe urls which are not on the whitelist. Although the number of urls frequented by users have been speculated to be comparatively limited, an alert will also be shown if a user attempts to upload sensitive personal identifiable information on a new legitimate url.

There exists the potential for the user to grow an insecure habit of quickly disregarding the alert any time he/she wants to upload sensitive information to a new url by executing actions that will allow and make the warning alert go away.

Heuristic Based Detection

The heuristic based approach is a method of detecting and preventing phishing attacks based on the investigation of the contents of phishing emails and urls to assess its validity. Initially, heuristic based approaches are meant to supplement the blacklist approach, because the blacklist approach appears to be unsuccessful in detecting new zero day phishing campaigns that have not been manually added to the blacklist after verification of legitimacy.

After careful review of relevant literatures, this approach can be subdivided into four major categories [21].

a. Rule Based Approach
b. Search Engine Assisted Approach
c. Content Similarity Approach
d. Visual Similarity Approach.

- Rule Based Approach
  This is the easiest type of heuristic detection to combat phishing attacks. To check for the validity of an email or url, the rule based approach employs a set of static rules. More often, these rules are designed based on the current trends of phishing attacks and manually designed for use in detecting phishing attacks.
  PhishCatch, a phishing email detection approach focused on heuristics was suggested by [28]. In order to check the validity of emails, PhishCatch depended on a weighted average performance from a series of rules, the rules centered on

the features of urls in the email, email content and PhishCatch's personalized blacklist [28].

PhishCatch's assessment of a 4804 email dataset containing 3710 phishing emails, revealed that PhishCatch could identify 80% of the present phishing emails with a false positive rate of approximately 1%. Yu et al. [28] indicated that urls that were not present in PhishCatch's blacklist could be identified by PhishCatch.

Cook et al. [29] suggested the stateless filter Phishwish for detecting phishing attacks. In order to determine the legitimacy of a website, Phishwish used a score derived from a weighted binary outcome average from each of its rules. If a website is above a certain threshold, the website is considered to be a phishing website. Phishwish was evaluated with a data set of 117 unique emails which contained 81 phishing emails, it revealed that the spectrum of legitimate and illegitimate emails was sufficiently distinct to allow Phishwish to distinguish phishing emails with high precision even though the rules are equal in weight. Nevertheless, Phishwish rules relied on HTML emails formats, and emails with a large amount of advertisements and urls could increase the likelihood of false positive scores.

- Search Engine Assisted Approach

It is often discovered in relation to phishing websites, that there exists a visual similarity to a legitimate website. This discovery is realistic as a phishing website aims to trick unsuspecting users into believing that he/she is on the legitimate website. This approach takes advantage of this discovery to create a detection mechanism by extracting keywords from suspected illegitimate websites and querying the keywords on a powerful search engine to evaluate the authenticity of the website. An anti-phishing toolbar used to identify phishing websites using the page content and a trustworthy search engine named CANTINA was created by Zhang et al. [30]. To classify the most relevant words on a webpage, CANTINA employed the TF-IDF algorithm to recognise the words which were used to create a lexical signature.

The approach suggested by Zhang et al. [30] suggests the use of CANTINA, a content-based strategy for the detection of phishing websites by means of TF-IDF measures. Using TF-IDF, CANTINA determines whether a website is legitimate or illegitimate. By counting its frequency, TD-IDF produces weights that determine the word context.

CHUENCHUJIT [21] describes CANTINA working process as:

1. For a given website, calculate the TF-IDF
2. Take the top 5 TF-IDF words and apply them to the url to locate the lexical signature
3. Send the lexical signature into a search engine.

After sending the lexical signature into the search engine, if the results of the N tops search result from the search engine, the website is deemed a legitimate website, otherwise it is a phishing website.

For the purpose of the experiment, N was set to be 30. Although if the search engine shows zero information, the website would be marked as a phishing website, this was the biggest inconvenience encountered in using this technique as this caused the rate of false positives to soar.

Dunlop et al. [31] proposed a plug-in for browsers that could use the optical character recognition (OCR) platform and a search engine to identify phishing websites, this browser plugin was named GoldPhish.

GoldPhish employed OCR technology to translate texts and upload them into a search engine from the screenshot of the website. The website is legitimate if the domain of the website is part of the top results. Average accuracy rating of 98%, a false positive rating of 0%, and a false negative rating of 2% were achieved while evaluating against 100 phishing and 100 legitimate websites.

GoldPhish's key constraints were the delays encountered while rendering the web pages from OCR and search engine lookups, susceptibility to exploitation via PageRank manipulation attacks and poor webpage efficiency with inadequate OCR texts.

- Content Similarity Approach

As the name implies, the method of phishing detection is based on analysing the content of a website. There exist similarities between the search engine approach and content similarity approach as they both leveraged the discovery that phishing websites were designed to be relatively similar to the legitimate websites and often make use of materials from the legitimate websites.

The identification of phishing websites based on visual similarities of contents between phishing and legitimate websites was a promising approach suggested by [32]. Based on the content similarities, this method breaks the website into major visual block regions. Three metrics then determine the visual similarities between the phishing and illegitimate web pages: similarity of blocks, similarity of structure and similarity of style. If any of these three metrics have a value greater than a predefined threshold, the webpage is considered to be a phishing webpage [32]. An experiment which contained 8 phishing webpages and 320 official bank webpages was carried out by researchers, the findings resulted in 100% true positive rate & 1.25% false positive rate. While its findings were remarkable, the high plasticity of web page structure caused this approach to be unreliable.

DomAntiPhish, an anti-phishing tool that exploited the similarity in the design of websites to detect the user entering his/her sensitive details on an illegitimate website was created by Rosiello et al. [33]. DomAntoPhish was an extension to an anti-phishing system called AntiPhish, which alerted the user when entering his/her sensitive details on an unknown website. DomAntiPhish compared the DOM object of the suspected website to the saved one and only raised a warning if the structural resemblance of the website is suspicious.

- Visual Similarity Approach

This approach to phishing detection computes the correlation between a suspicious website and a registry of legitimate website features which include screenshots, logos, favicons, icons and document-oriented models. If the similarity score is calculated and the result is above a threshold, it suggests that the suspicious website is imitating a legitimate website. This approach is beneficial as cyber criminals often clone legitimate websites in order to trick unsuspecting users into parting away with their sensitive personal data. Nonetheless, this technique is far from infallible as the phisher can easily circumvent it, if any usual element is slightly changed without impacting the overall appearance of the imitated page.

A phishing identification mechanism based on Favicon recognition was suggested by [34] Favicon is a trademark image linked to a specific website. In order to confuse unsuspecting users, phishers forge the favicon of a legitimate website to put on the phishing website.

On a suspected website, the suggested mechanism identifies the favicon and conducts a visual similarity comparison of the favicon against a collection of favicons from legitimate websites. The suggested approach was evaluated against a data collected comprising 3,642 illegitimate websites and 19,585 genuine websites and revealed a true positive rate of 99.6% and a false positive rate of 2.79%.

Machine Learning Based Detection

In recent years, the detection of phishing emails, urls and messages has primarily become a matter of classification. Researchers and scientists have turned their focus into a machine learning approach for the detection and classification of phishing attacks.

A host of different techniques for machine learning have been investigated and employed including support vector machines (SVM), neural networks and decision trees. This techniques have been used to develop a number of anti-phishing technologies to recognise inconsistencies between the structure of the websites, their HTTP transfers, suspected validity of the website and phishing emails [35].

Machine learning approaches to detecting phishing attacks are far more versatile and can recognise the modifications in phishing websites that escape detection in other available approaches for detection. If supplied with ample training data, this approach possesses the capability to combat against zero day phishing attacks. This approach is impressive in its detection and classification, plus it is highly dependent on the scale, value and precision of the training data set to achieve optimum precision [36].

In 2011, the 47 features of the email which were used in classifying phishing emails were included in a study carried out by Khonji and others, each feature was briefly explained and they covered all the structures of an email [37].

In terms of precision, Azad had concentrated on evaluating many existing algorithms such as Naïve Bayes, SVM classifiers and Logistic Regression. He employed bag of words and augmented bag of words model for his evaluation. The evaluation indicated the SVM with linear kernel & Naïve Bayes leading other classifiers in terms of high results, suggesting an accurate rate of 95%, as they skipped 10 and 2.66% of the phishing emails respectively.

In comparison, the SVM showed similar outcome with lower features in contrast with the Naïve Bayes and Logistic Regression. Concurrently, the linear SVM was evaluated with additional features deleted which resulted in a poor detection rate as 5.86% of phishing emails were misclassified, signalling that additional features increases the precision of the result. In conclusion, the analysis reveals that linear SVM is effective for phishing email detection before they get into user's inbox [38].

Dewan et al. [39] present a model to detect spear-phishing emails using social and stylometric features to detect spear-phishing emails. The social features were extracted from the LinkedIn profile of employees form 14 different international organizations and a real world dataset from symantec enterprise email scanning service was used. Three classification tasks were carried out. In two out of the three tasks, it was discovered that the social features extracted from the LinkedIn profiles of various employees did little or no help in helping to classify whether a mail was spear-phishing or not. An average of 97.3% was achieved when the social features and stylometric features were implemented. However, without the social features the classification tasks achieved an average of 98.23%, the analysis revealed that social features are ineffective for spear-phishing email detection.

A technique suggested by James and George used a feedforward neural network and extraction of features in email headers and html body to identify whether an email was legitimate or not. Using 5 secret neurons, their suggested algorithm was tested on 18 features. A 173.55 ms training is required for the algorithm before applying it, and a single email is tested in 0.00068 ms. With the rize in the number of neurons, the time spent will increase though it is still considered as minimal. As for the results, 98.72% precision and 0.01 learning rate were exceptionally reliable [40].

Almomani found a new model in 2013, that showed superior outcomes compared to other approaches in terms of precision, true negatives, sensitivity, true positives, F-measurement and overall accuracy. Furthermore the model demonstrated reliability in its online prediction of values of emails and long term use of stored memory. The model proposed by Almomani named Phishing Dynamic Emerging Neural Fuzzy System (PDENF) was developed to predict phishing emails and aid in the detection of zero day phishing email attacks [41].

Kathirvalavakumar et al. [42] suggested a model that uses the neural network approach of exclusion pruning to classify an email as legitimate or not. 18 features reflecting the diverse facets of emails are derived from the suggested model. A dataset of 4000 phishing and legitimate emails was used to test the suggested model. The test resulted in a high accuracy rating of 99.9% for classification of emails. Although the generalization problem usually encountered when using neural networks as the center of detection was the main drawback to this approach.

EmailProfiler, a headerand stylometric email feature system proposed by [43], to defend users against spear-phishing attacks. The system identifies whether the received email originates from the sender stated in the metadata of the particular email, the system creates a behavioral profile for each sender and compares incoming mails with the existing profiles while also creating stylometric features which include lexical, syntactic and structural features. Both features from the email metadata and stylometric information are evaluated on a support vector machine classifier using a data set which comprises 20 volunteers sending both legitimate and spear-phishing emails. The experiments achieved an accuracy of 98%, due to the small number of data used the system is not ready to be deployed for spear-phishing detection.

Smadis suggested a detection mechanism for phishing emails that uses 23 features derived from email header and body on a J48 decision tree classifier. The features used in the proposed model include the message domain, sender domain, message type, number of links and URL features in the link. The model for classification was evaluated using a data set of 4559 legitimate and illegitimate emails with tenfold cross validation, the evaluation resulted in a 98.11% accuracy and 0.53% false positive rate [44].

There is no general consensus on any one of the features as the best for classifying and detecting spear-phishing emails. The research works that were reviewed recorded a very high accuracy rate with a relatively low false positive rate. Majority of the features used in this approach for detection was frequently derived from email headers, message content, and also external information such as from social media profiles. Each research studied proposed their own set of features via an evaluation process based on a dataset consisting of both legitimate and spear-phishing emails.

3 Methodology

The proposed spear-phishing email classification method adopted the Knowledge Discovery in Database (KDD) in creating an intelligent classifier for emails, that is able to classify a new email as either a legitimate or a spear-phishing email.

This proposed spear-phishing email classifier applies the iterative measures of KDD to define and extract useful features from a training dataset, the extracted features are then implemented on the classification models to determine the best classifier.

3.1 KDD

KDD refers to the general process of discovering meaningful knowledge from data. It requires the evaluation and likely analysis of patterns in order to determine what qualifies as knowledge. It also involves the option of encoding schemes, sampling, preprocessing and estimates of the data prior to the data mining stage [45] (Fig. 1).

The process of discovering & interpreting patterns from data calls for the repeated application of a set steps. This steps include:

1. Understanding the problem: This includes the application domain, having the relevant prior knowledge concerning the problem and understanding what the user aims to achieve.

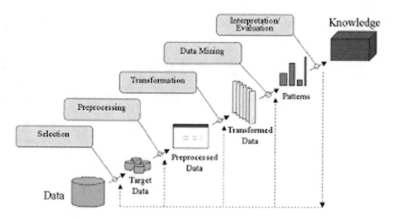

Fig. 1 Knowledge discovery in database methodology. *Source* KDD [45]

2. Producing target dataset which focuses on the variables in which the users aims to discover knowledge.
3. Cleaning and Preprocessing of the target dataset in order to remove noise and ensure that the necessary information needed are present in the dataset.
4. Reduction & Projection of Data to find the useful features needed to represent the data based on the end goal.
5. Data mining task selection: Deciding the end goal of the process i.e. classification, regression, clustering etc.
6. Pick the algorithm for data mining: Selecting the methods to probe for patterns in the data.
7. Data mining: Probing for patterns in the training dataset.
8. Interpreting the mined patterns.
9. Knowledge is discovered and then presented in a useful and easy to use information.

The adoption of the KDD methodology assisted in the creation of the methodology and process flow used for the identification and binary classification of spear-phishing emails.

The framework diagram below displays the process flow of activities taken to achieve the objectives of this research paper. This illustrates the flow of execution from the compilation of the dataset to the delivery of the findings (Fig. 2).

3.2 Data Collection

The quality of data is of utmost importance when dealing with machine learning as it directly influences the performance of the machine learning model in which it is implemented on [46]. The initial step required in developing the proposed model for

Fig. 2 Process flow diagram

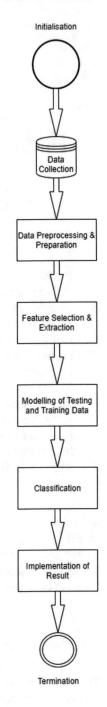

identifying and binary classification of spear-phishing emails is dividing the training dataset which comprises both spear-phishing and legitimate emails. For the purpose of this research, the Enron email dataset was obtained from dataworld containing approximately 500,000 emails generated by Enron Corporation employees. This dataset was obtained during an investigation into Enron's collapse by the Federal Energy Regulatory Commission. The Enron email dataset has no attachment, and a few email messages were deleted per request from affected parties in Enron. In a bid to consider attacks coming from third parties, a copy of the emails in the dataset were simulated from @enron.com to yahoo ymail platform (@ymail.com) which increases the emails in the dataset to over a million emails. About 6804 emails were selected for the training purposes, which provided the opportunity to have an equal number of phishing & legitimate emails.

3.3 Data Preprocessing

As stated earlier, the quality of the data used on a machine learning algorithm is of high importance. The data preprocessing stage involves strategies to transform the raw dataset available in order to generate quality data, specifically including standardization, data creation, formatting, reduction and discretization of the dataset [47]. This is done to ensure that the training dataset is complete, error-free and up to standard, to yield optimal and accurate results when passed through the machine learning algorithms.

This stage required a great deal of effort and time to ensure that discrepancies in the training dataset were handled before the dataset can be applied on the machine learning algorithms.

3.4 Feature Extraction

At this stage, feature extraction is done to decrease the amount of features used on the proposed model, this process is carried out to speed up the classification of the spear-phishing emails.

There are two phases considered in this stage, the first phase selects features to be drawn out from texts and headers of each email, these features define various properties of the email. The second phase involves selecting the most efficient features from the extracted set in the first phase.

For the purpose of this research paper the features to be used are selected from two sources:

1.   Email Header: Email headers contain sensitive information that assists in identifying where the email originates from. This information can include the location of the email origin, IP address block, email address, display name and content

type. This research paper will be considering four features from the email header namely; Message ID, email address (FROM), username/display name & label (Ham or Spam)

2. Email Content: The content of the email communication is designed to include substantive details intended for the receiver to infer the actual nature of the email. Features such as HTML body feature, subject feature, are part of the features to be considered in this research paper.

The emails in the dataset is processed to extract features to be used, the process includes:

- Tokenization: A common task in Natural Language Processing (NLP), which is a means of dividing a piece of text into smaller units called tokens.
- Stop words removal: Omitting common vocabulary that seems to be of little or no meaning from the tokens such as "myself", "me", "that", "I've"....etc. This process is carried out to reduce the similarity between emails in the dataset. It enhances and improves the efficiency of the proposed model in it's implementation.
- Word Weighting: This step is taken to help identify the words with the highest frequency in the dataset, this is done to indicate the importance of these terms in identifying phishing emails.

After the feature extraction from the email header and email content is carried out, the results are concatenated together to give a new finalised dataset that is ready to be modelled and evaluated.

3.5   Modelling and Evaluation

Modelling of a dataset involves dividing the dataset into two distinguishable sets i.e. training and testing data. For the purpose of this research paper the training and testing were divided into ratio 70% and 30% respectively, owing to the large size of the available dataset. This ratio assists the model in producing incredibly detailed results by taking several related inputs for the proposed model to recognise the unique underlying patterns and using smaller samples to validate the ability to recognise patterns. The modelled dataset will be parsed through Decision Tree, Support Vector Machine, Logistic Regression, Random Forest and K-Nearest Neighbour classifiers using python language. Libraries such as Numpy, Matplotlib, Natural language toolkit, Pandas, Scikit-Learn and Seaborn.

In a bid to access the overall success of the proposed model, a variety of metrics will be used. Evaluation process is carried out to define the importance or significance of a body of work. In this case, identifying phishing emails using machine learning. A few metrics was selected because of its suitability for binary classification activities, such as precision, recall, accuracy score, confusion matrix etc.

3.6 Classification Model Building

For the purpose of this research, to train and evaluate the accuracy of spear-phishing emails detection with the group features discussed earlier, five supervised classification algorithms were selected.

The rationale behind the selection of these algorithms is the various training strategies used to explore the rules and processes of learning and testing. The following selected algorithms are considered as popular algorithms.

3.6.1 Random Forest Algorithm

This is a supervised learning algorithm that consists of a broad number of individual decision trees, and is suitable for performing both regression and classification tasks. A combination of tree predictors whereby each tree in the forest relies on the independently sampled random vector values, the generalization error for forest trees converges to a point as the number of trees in the forest increases. A generalization error in a forest of tree classifiers is dependant on the intensity and relation between the individual trees in the forest [48].

The advantages of Random Forest include increased precision, ability to handle missing information, not affected by noise, fits well for handling continuous and categorical variables, effectively manages nonlinear parameters, and is resilient to outliers. These benefits work simultaneously to ensure that when applying the Random Forest algorithm it delivers reliable, consistent, secure and resilient outcomes across the board.

3.6.2 Support Vector Machine (SVM)

This is also a supervised learning algorithm which in addition to regression tasks, is often used to carry out classification tasks. SVM plots each data item as an n-dimensional point space (n is the number of features in the training set of each sample) and the main aim of the algorithm is to find the best hyper-plane that separates the two groups. By translating into a higher dimensional space (using a kernel function) where a separate hyperspace exists, SVM is able to classify non linearly separable data. It's precision and ability to separate non linearly separable data are the advantages of SVM. Although, it is memory-intensive and difficult to explain [49].

3.6.3 Logistic Regression

Logistic regression is a supervised learning algorithm that is commonly employed ny staticians and researchers as a traditional statistical and data mining method for the estimation, recognition and classification of binary and proportional response dataset [50].

Logistic Regression can be used to handle classification issues, it offers the binomial results as it presents the probability of an event occurring or not occurring in terms of 0's and 1's which is dependent on the values of the input variable. It's advantages include predictive performance, flexibility of implementation, eaze of regularization and efficiency based on training.

3.6.4 Decision Tree

A supervised learning technique used to solve classification and regression problems by continuously separating data depending on a given condition with the decisions in the leaves of the tree and the data in the nodes of the tree are divided. It is similar to a tree graph where grouping starts before completing the goal from the root node to the leaf node. It is commonly used, resilient to noisy information, and a functional technique for learning disjunctive expressions [45].

The key benefit of decision trees are their ease of describing, understanding and taking into consideration the relationships between features and it's interactions. Nevertheless, it does not encourage online learning and requires the tree to be reconstructed every time new samples occur [49].

3.6.5 K-Nearest Neighbour (KNN)

An efficient technique of supervised learning for many computation problems. It's algorithm suggests that identical objects occur in close proximity to each other. KNN operates by finding the distances connecting the data samples and a query, choosing the given number of examples (k) nearest to the query, the voting for the most common label in case of classification or averaging the label in case of regression [51].

In KNN, the value of K depends on the scale of the dataset and the form of problem for classification. The bulk of the K-Nearest Neighbour would be considered as a class for the test data after gathering the K-Nearest Neighbours [52] (Fig. 3).

3.7 Performance Metrics

After implementing different classification techniques on the proposed model for detecting phishing email, the next step is to apply a set of evaluation metrics to check for the efficiency of each classification technique. This research paper will check the efficiency of the classification techniques using the following performance metrics:

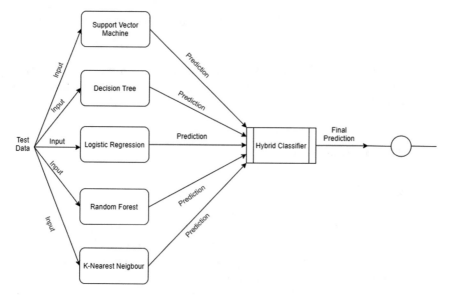

Fig. 3 Proposed system for detection of spear-phishing emails

- Accuracy Score: This is the ratio between the number of accurate predictions and the number of input samples in general, popularly regarded as the simplest metric for evaluating in machine learning. It clearly defines the section of properly classified values which represents the precision of the classifier. It is ideal to use the accuracy score metric when the number of samples belonging to each class is matched or equivalent.
- True Positive Rate (TP): This refers to the percentage of phishing emails in the training dataset that the algorithm accurately classified i.e. the amount of times the actual positive values are equivalent to the predicted positive value.
- True Negative Rate (TN): This refers to the percentage of legitimate emails in the training dataset that the algorithm accurately classified i.e. the amount of times the actual negative values are equivalent to the predicted negative value.
- False Positive Rate (FP): This refers to the percentage of legitimate emails in the training dataset that the algorithm inaccurately classified i.e. the amount of times the negative values are inaccurately predicted by the algorithm as positive value.
- False Negative Rate (FN): This refers to the percentage of phishing emails in the training dataset that the algorithm inaccurately classified i.e. the amount of times the positive values are incorrectly predicted by the algorithm as negative values.
- Precision: This metric refers to the model's ability to correctly classify value i.e. what proportion of emails the classifier marks as phishing are actually phishing emails.
- Recall: This metric refers to the model's ability to predict value i.e. what proportion of phishing emails the classifier marks as phishing emails.

- F1 score: This is also referred to as F-measure and is defined by the harmonic mean of precision and recall, it illustrates how efficient and durable the classifier is.
- Receiver Operating Characteristic (ROC) Area: This is a performance metric that displays a binary classifier's accuracy by plotting TP against FP at different threshold values.

4 Design Framework

4.1 Model Framework

Beginning with the dataset that will later be simulated to present emails from within and outside Enron Corporation (Data Collection). Data processing is done to ensure the dataset is error free and ready to be parsed. Feature selection is carried out on the dataset to extract the required features that will be used to train and test the classifiers, and also speed up the classification process, thereby achieving a new dataset to be rationed 70:30% for the training and testing on the classifiers (Modelling).

The model is trained using 5 different classifiers and a performance evaluating is carried out on each classifier to help determine their level of accuracy. The best two classifiers from the performance evaluation are combined to achieve optimal performance and predictions (Fig. 4).

4.2 Sequence Diagram

The sequence diagram below for the proposed model gives a visual representation of how the model works to determine whether an email is legitimate or not on an email gateway i.e. The moment the user receives the incoming emails, how the features of the emails are extracted and evaluated before a prediction is given (Fig. 5).

4.3 Requirement Specification

There are certain requirements essential for the smooth implementation and running of the proposed model, they are divided into two separate sub-sections:

1. Software & Hardware Specification: The hardware and software specification required for this research are highlighted in the table.
2. Functional & Non-Functional Requirements: System requirements in machine learning are often classified as functional requirements (FRs) and Non-functional requirements (NFRs). Kurtanović & Maalej [53] mentioned that

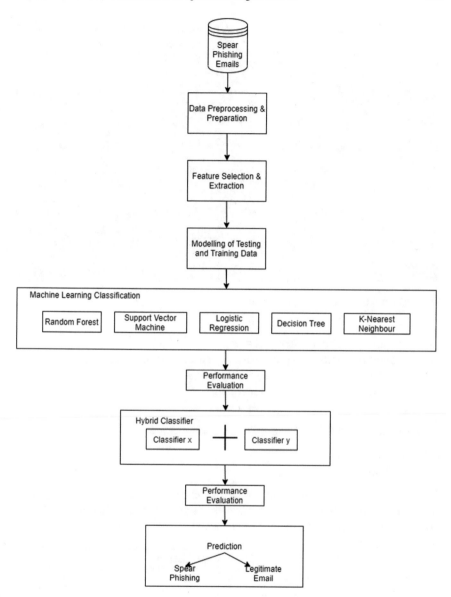

Fig. 4 Architectural design of proposed system

FRs refer to the functionality of a specified system i.e. the job description of that particular system that has been specified to the user. Meanwhile NFRs refers to the properties and constraints of the specified system. The FRs of the proposed model include:

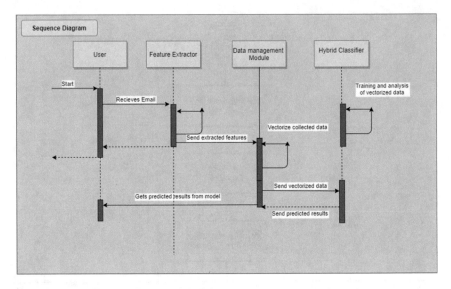

Fig. 5 Sequence diagram of proposed model

(a)   Capability to accept emails as data
(b)   Capability to carry out pre-processing task on the accepted data
(c)   Capability to train all five Classifiers (RF, K-NN, DT, SVM and LR)
(d)   Classifying the emails as spam or ham
(e)   Carry out evaluation using predefined metrics.

Meanwhile the NFRs of the proposed system include:

(a)   Speed
(b)   Reliability.

These NFRs are relevant if the proposed model is implemented to sensitive systems where timing is essential to the effectiveness of an attack, therefore the model will need to be relied on to stop the impending attack.

	Requirements	Specification
Software	Web client	Chrome
	Programming language	Python
	IDE	Anaconda Navigator, Jupyter notebook
	Operating system	Linux, windows or MacOs
Hardware	Memory	4 GB minimum
	CPU	Intel Core i3 or higher

5 Implementation

5.1 Programming Language

This work was done and written with the python programming language. Python is a simple, high level, general purpose and object-oriented programming language that places strong emphasis on code readability. For this research paper, python programming language is the best fit due to the abundance of frameworks and libraries that render prototype creation focused on machine learning and artificial intelligence.

5.2 Library

A lot of powerful machine learning libraries and frameworks are available on the python platform such as Tensorflow, Sklearn, Numpy, Pytorch etc.

As seen in Fig. 6, the necessary libraries required for the implementation of the proposed model are imported.

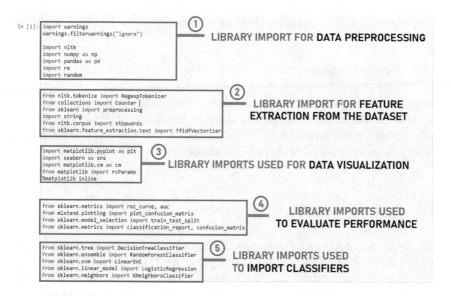

Fig. 6 Importing libraries necessary for the implementation

5.3 Tools

The following tools listed below were used in the developmental process of the proposed model.

5.3.1 IDE

This is an integrated development environment, it is a software application that offers extensive facilities alongside a graphical user interface, where computer programmers can create software packages and programs.

The IDE used for the research paper was the Anaconda Navigator and PyCharm, which are specifically designed for python programming.

5.3.2 Anaconda Navigator

An application software compatible with windows, MacOs and Linux, it is part of the Anaconda R distribution, it provides a flexible way to navigate conda packages, install python libraries and use third-party data science libraries. It provides a platform to carry out multiple data science projects simultaneously.

5.3.3 Jupyter Notebook

An open source web application that provides a platform for creating and distributing documents which entails equation, visualizations, live codes and text documents. This web application was used to better visualize graphs, charts, numerical simulations, mathematical modelling and to visualize the evaluation of the model.

5.4 Implementing Proposed System

Data Importation: After data collection, the dataset that would be employed to carry out the aims of the research paper was imported using the python library (Panda) into the python file. Panda offers a simple, versatile and expressive data structure developed to work with structured and time series data such as the dataset to be employed.

The dataset was stored in the csv (comma separated value) format, so as to be easily read and edited using python. The dataset was labelled using the read.csv function (data=pd.read_csv("dataset.csv")).

5.4.1 Data Visualization

Due to the large nature of the dataset, it is important to try and understand the data by visualizing it. Such that, patterns, trends, and correlation in the available dataset are easily taken note of. Some of the visualization libraries used in this research paper include matplotlib, Seaborn, Pylot which visualized the data in forms such as histogram and pie chart.

5.4.2 Data Balancing

There exist several aspects that might influence the performance of a machine learning algorithm and data balancing is one of them. Class imbalance where a training data belonging to a particular class heavenly outnumbers the other class is an example of an imbalanced data which will lead to the machine learning algorithm making inaccurate predictions. Therefore, the dataset used for this research employed the SMOTE method to balance the training dataset evenly.

5.4.3 Data Splitting

In order to achieve an impartial assessment of predicted results, data splitting is necessary. Model evaluation and estimation is one of the main facets of machine learning, when assessing the model's predictive efficiency, it is crucial that the process is unbiased. This process involves splitting the dataset into training and test data.

The dataset was divided using the Sklearn.model_selection train test split (), the ratio considered for the implementation process was 70:20 with the training dataset accounting for 70% while the testing data accounted for 20%.

5.4.4 Feature Scaling

It improves the performance of the machine learning model used for the prediction in this research. The data set used in the implementation of the proposed system had multiple features which required pre-processing (feature scaling) to normalize the data within a particular range. This was done using the Standard Sealer class from the Sklearn.preprocessing library on the training and testing dataset.

5.4.5 Model Building

The first step here is determining what classifier to be used, in the case of this research 5 classifier model were deployed in the implementation of the proposed system for this research paper. The five classifier packages were imported using the Sklearn ensemble import " Classifier", defined and fit into the model before evaluation using (x_train, y_train).

5.4.6 Evaluation of Classifiers

In order to evaluate the performance of the classifiers used for classification tasks, it is important to have successfully trained the classifiers. By using different metrics for the performance evaluation of each of the classifiers, the overall efficiency and effectiveness of each classifier was assessed which in turn assist to improve overall positive power of the proposed system. The confusion matrix (y_test, y_pred), accuracy_score (y_test, y_pred) and classification report (y_test, y_pred) were used in evaluating the outcomes of the binary classification task.

6 Evaluation

This section aims to evaluate the results obtained for each of the five classifiers and the feasibility of the proposed model. The Enron email dataset was used in carrying out this experiment, as it contains spear-phishing emails. With the aim of detecting spear-phishing emails, the results obtained were analysed to answer the research question "**Can the detection and binary classification of spear-phishing emails in organization be improved by using a hybrid machine learning classifier?**".

To answer the above stated question, five different machine learning classifiers were evaluated, and the results were compared using the following performance metrics: accuracy score, recall, confusion matrix, F1 score and precision. After evaluation, the two machine learning classifiers with the best results were combined to create a hybrid classifier. The result from the hybrid classifier is compared to that of the individual machine learning classifier to answer the above stated research question.

A brief explanation of the performance metrics used are:

- Confusion Matrix: A table that describes the performance of a classification model with some parameters. These parameters include True Positive (TP) which refer to emails that are classified as spear-phishing correctly, True Negative (TN) which refers to emails correctly classified as legitimate emails, False Positive (FP) which refers to emails that are incorrectly classified as spear-phishing, and False Negative (FN) which refers to emails that are incorrectly classified as legitimate emails.
- Accuracy: This simply refers to the proportion of correctly predicted observation to the number of total observations, often regarded as the most intuitive performance measure, it is defined using the mathematical expression (Accuracy = TP + TN/TP + FP + FN + TN).

- Precision: This simply refers to the proportion of positive observation correctly predicted against the total number of predicted positive observations. It is defined using the mathematical expression (Precision = TP/TP + FP).
- Recall: This is also known as Sensitivity, it refers to the percentage of positive observations correctly predicted to the total number of all observations in any given class. It is defined using the mathematical expression (Recall = TP/TP + FN).
- F1 Score: This takes into consideration both the false positives and false negatives. It refers to the weighted average of the precision and recall. It is very efficient in the case of an unbalanced class. It is defined using the mathematical expression (F1 Score = 2*(Recall*Precision)/(Recall + Precision)).

6.1 Experiment 1 Random Forest Classifier

The first classifier implemented was the Random Forest classifier using the Random Forest algorithm, this classifier attained an accuracy score of 97.65%, precision score of 97.65%, recall score of 97.6%, F1 score of 97.6%, and ROC of 97.6%. The recorded false negative of 22 and false positive of 26, i.e. false negatives refer to spear-phishing emails incorrectly classified as legitimate while false positives are legitimate emails incorrectly classified as spear-phishing. A low number is ideal for both FP and FN for an optimal algorithm.

6.2 Experiment 2 Logistic Regression Classifier

The second classifier implemented was the Logistic Regression classifier using the Logistic Regression algorithm, this classifier attained an accuracy score of 99.71%, precision score of 99.7%, recall score of 99.7%, F1 score of 99.71%, and ROC of 99.7%. The recorded false negative of 0 and false positive of 6, i.e. false negatives refer to spear-phishing emails incorrectly classified as legitimate while false positives are legitimate emails incorrectly classified as spear-phishing. A low number is ideal for both FP and FN for an optimal algorithm.

6.3 Experiment 3 Decision Tree Classifier

The third classifier implemented was the Decision Tree classifier using the Decision Tree algorithm, this classifier attained an accuracy score of 99.61%, precision score of 99.6%, recall score of 99.61%, F1 score of 99.61%, and ROC of 99.61%. The recorded false negative of 0 and false positive of 8, i.e. false negatives refer to spear-phishing emails incorrectly classified as legitimate while false positives are legitimate

246 P. F. Akinwale and H. Jahankhani

emails incorrectly classified as spear-phishing. A low number is ideal for both FP and FN for an optimal algorithm.

6.4 Experiment 4 Support Vector Machine Classifier

The fourth classifier implemented was the Support Vector Machine classifier using the Support Vector Machine algorithm, this classifier attained an accuracy score of 92.56%, precision score of 92.2%, recall score of 92.6%, F1 score of 92.5%, and ROC of 92.6%. The recorded false negative of 142 and false positive of 10, i.e. false negatives refer to spear-phishing emails incorrectly classified as legitimate while false positives are legitimate emails incorrectly classified as spear-phishing. A low number is ideal for both FP and FN for an optimal algorithm.

6.5 Experiment 5 K-Nearest Neighbour Classifier

The fifth classifier implemented was the K-Nearest Neighbour classifier using the K-Nearest Neighbour algorithm, this classifier attained an accuracy score of 94.52%, precision score of 94.68%, recall score of 94.54%, F1 score of 94.51%, and ROC of 94.54%. The recorded false negative of 89 and false positive of 23, i.e. false negatives refer to spear-phishing emails incorrectly classified as legitimate while false positives are legitimate emails incorrectly classified as spear-phishing. A low number is ideal for both FP and FN for an optimal algorithm.

For the above experiments, the two classifiers with the highest accuracy are the Logistic Regression and Decision Tree classifiers with an accuracy score of 99.71% and 99.61 respectively, the classifier with the lowest accuracy score was the Support Vector Machine classifier with an accuracy score of 92.56%. This is described in the table and Figs. 7 and 8.

```
Comparison of all algorithm results
+-----------------------------------+----------+
|               Model               | Accuracy |
+-----------------------------------+----------+
|       Random Forest Algorithm     |  0.9765  |
|   Logistic Regression Algorithm   |  0.9971  |
|       Decision Trees Algorithm    |  0.9961  |
| Support Vector Machine Algorithm  |  0.9256  |
|            KNN Algorithm          |  0.9452  |
+-----------------------------------+----------+
```

Fig. 7 Five classification technique comparison table

ROC CURVES OF ALGORITHMS

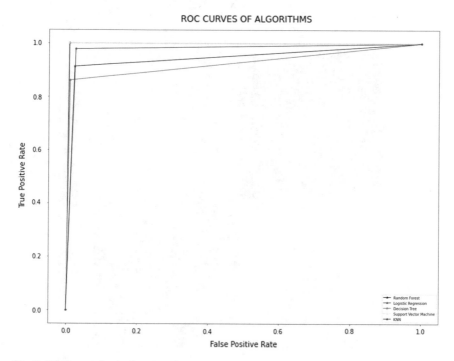

Fig. 8 ROC curve for the five classification techniques

6.6 Experiment 6 Hybrid Classifier

The hybrid classifier was implemented using the ensemble technique and multi_class taking the best two classifiers with the highest accuracy from the previous experiments. This hybrid classifier achieved an accuracy score of 98.8% with 0 false negatives and 4 false positives. The results are shown in the figures below showing the confusion matrix and summary table of the result (Figs. 9 and 10).

6.7 Discussions

To answer the research question "**Can the detection and binary classification of spear-phishing emails in organization be improved by using a hybrid machine learning classifier?**", this research paper took into consideration five machine learning classifiers, implemented, evaluated and compared the result before making an hybrid classifiers with the best two classifiers. It was discovered that the logistic regression and decision tree classifiers achieved the highest accuracy score during evaluation, out of the five machine learning classifiers used in this research. The two classifiers were combined to create a hybrid classifier. The table below shows a

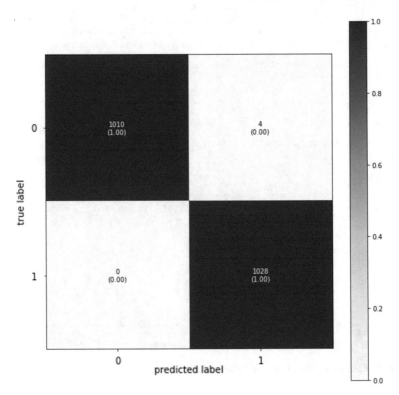

Fig. 9 Hybrid classifier confusion matrix

```
Precision, Recall, F1


                    precision    recall  f1-score   support

               0      1.0000     0.9961    0.9980      1014
               1      0.9961     1.0000    0.9981      1028

        accuracy                           0.9980      2042
       macro avg      0.9981     0.9980    0.9980      2042
    weighted avg      0.9980     0.9980    0.9980      2042
```

Fig. 10 Hybrid classifier result table

description of the results obtained from the test run with the different classifiers on the Enron email dataset (Table 1).

As observed from the table above, the hybrid classifier marginally outperforms the Logistic Regression and Decision Tree classifier with a 99.8% accuracy score compared to logistic regression's 99.7% and decision tree's 99.61%.

Table 1 Comparison table (hybrid and the five classification techniques)

Algorithm	Accuracy (%)	Precision (%)	Recall (%)	F1 score (%)	FP	FN
Random forest	97.65	97.65	97.6	97.6	26	22
Logistic regression	99.71	99.7	99.7	99.71	6	0
Decision tree	99.61	99.6	99.61	99.61	8	0
Support vector Machine	92.56	92.2	92.6	92.5	10	142
K-Nearest neighbour	94.52	94.68	94.54	94.51	23	89
Hybrid classifier	99.8	99.81	99.8	99.8	4	0

The remaining three classifiers performed much worse in terms of accuracy with the random forest achieving 97.65%, k-nearest neighbour achieving 94.52% and the worst of the lot support vector machine achieving the lowest with a 92.56% accuracy score.

Both the hybrid classifier and the logistic regression classifiers are suitable for the binary classification task, as they both achieved a high degree of accuracy. The difference recorded in accuracy between the hybrid and logistic regression classifier is 0.1%, it can be debated to not make the hybrid classifier an ideal solution as it takes time to train nor does it justify the amount of computational resources needed to train the classifier. Although it is a valid point, it is still necessary to remember that when it comes to email security, the smallest accuracy percentage can be the determining factor between an attacker effectively downloading malware on a corporate network or defrauding a target (user or organization) of millions in financial value. Therefore, it is ideal as the classifier only needs to be trained once, the higher computational resources required by the hybrid classifier is a small price to pay for increased security.

Clement [54] mentioned in her report that approximately 306.4 billion emails were sent and received in the year 2020, the number is projected to increase to 361.6 billion by 2024 creating a volume of 306.4 million emails when evaluating 0.1% of 306.4 billion. 306.4 million spear-phishing emails attacks can theoretically be eliminated if an email gateway similar to the hybrid classifier is introduced compared to the logistic regression classifier. Therefore the 0.1% is not as negligible as it initially seemed.

The hybrid classifier appears to be the cream of the crop of the mentioned classifiers in this research paper, it is not withouts its drawbacks. The time taken to train and test the classifier, the limited volume of data used for both testing and training, and the burden that this classifier places on the computational power are some of its current drawbacks.

Finally, having carefully reviewed the findings of the experiments and thoroughly analyzed them, it can be said that the hybrid classifier can potentially enhance the identification and binary classification of spear-phishing emails, thus answering the research question proposed in this research.

7 Conclusion and Future Works

Spear-phishing emails have become a common problem in recent years, it is a type of phishing attack in which the attacker employs social engineering techniques in a coordinated attempt, mostly for malicious purposes, to steal confidential information such as account credentials, financial information from a particular victim or install malicious programs on an organization network.

This research is aimed at answering the research question **"Can the detection and binary classification of spear-phishing emails in organization be improved by using a hybrid machine learning classifier?"**. The use of five supervised machine learning algorithms namely Random Forest, Logistic Regression, Decision Tree, Support Vector Machine and K-Nearest Neighbour were used to detect spear-phishing emails. In this thesis, a hybrid classifier is presented, the best two classifiers (Logistic Regression, Decision Tree) from the five supervised machine learning algorithms were merged to achieve a good accuracy score of 99.8% for the detection of spear-phishing emails, indicating the effectiveness of the proposed model. The model was built using an intelligent pre-processing and features selection phase that extracts a set of features from the email header and the email content.

Comparing the level of accuracy of the hybrid classifier with five other machine learning algorithms addressed the above stated research question. And overall, the hybrid classifier proved to be more precise. Therefore, the research question was answered satisfactorily.

However, this research paper was restricted by the limited amount of sample data available as well as a sufficiently strong computational resource to make the proposed system as optimal as possible. Furthermore, the short period of time available for this paper to thoroughly grasp and understand topics related to machine learning, python programming and analysis methodologies further restricted the depth of the research.

To investigate whether this will improve the overall accuracy of the proposed model, future researchers should consider using a much larger dataset, and also consider emails that contain attachments and urls as the Enron email dataset does not include attachments and urls.

There already exists space for improvement of the proposed system, a new emerging threat where attackers are using attachments to send the text of an email, a new technique where both text from the email body and attachments are extracted can be implemented.

References

1. Al-Saaidah SA (2017) Detecting phishing emails using machine learning techniques, s.l.: s.n
2. APWG (2013) About us. https://apwg.org/about-us/. Accessed 1 Nov 2020
3. APWG (2020) Phishing activity trends report 2nd quater 2020, s.l.: APWG

4. Kevin T (2018) 18.5 million websites infected with Malware at any time. https://www.securi tyweek.com/185-million-websites-infected-malware-any-time#:~:text=%22We%20went%20from%20about%200.8,malware%20at%20any%20given%20time. Accessed 1 Nov 2020
5. Google (2020) Google safe browsing. https://transparencyreport.google.com/safe-browsing/ overview?unsafe=dataset:1;series:malwareDetected,phishingDetected;start:1148194800000; end:1604214000000&lu=unsafehttps://transparencyreport.google.com/safe-browsing/ove rview?unsafe=dataset:1;series:malwareDetected,phishingDetected;start:1148194800000;end: 1604214000000&lu=unsafe. Accessed 1 Nov 2020
6. CISCO (2020) What is phishing. https://www.cisco.com/c/en/us/products/security/email-sec urity/what-is-phishing.html. Accessed 2 Nov 2020
7. Irwin L (2020) The 5 most common types of phishing attack. https://www.itgoverna nce.eu/blog/en/the-5-most-common-types-of-phishing-attack#:~:text=The%205%20most%20common%20types%20of%20phishing%20attack.,4.%20Smishing%20and%20vishing.%205%205.%20Angler%20phishing. Accessed 3 Nov 2020
8. NORTON (2018) What is smishing. https://us.norton.com/internetsecurity-emerging-threats-what-is-smishing.html. Accessed 3 Nov 2020
9. Best A (2020) What is the definition of SMiShing?. https://inspiredelearning.com/blog/what-is-the-definition-of-smishing/#:~:text=What%20is%20a%20SMiShing%20Attack%3F%20SMiShing%20is%20the,malicious%20website%2C%20or%20calling%20a%20fraudulent%20phone%20number. Accessed 3 November 2020
10. FraudWatch (2019) What is vishing? voice phishing scams explained & how to prevent them. https://fraudwatchinternational.com/vishing/what-is-vishing/. Accessed 3 Nov 2020
11. Centeno L (2020) 8 types of phishing attacks you should know about. https://www.makeuseof.com/types-of-phishing-attack/. Accessed 3 November 2020
12. Frauenstein ED (2013) A framework to mitigate phishing threats. Nelson Mandela Metropolitan University, South Africa
13. Rader M, Rahman SS (2015) Exploring historical and emerging phishing techniques and mitigating the associated security risks. Int J Netw Secur Appl (IJNSA) 5
14. DeFino S, Greenblatt L (2012) Official certified ethical hacker review guide: for version 7.1—EC-Council Certified Ethical Hacker (Ceh), 1st edn. Course Technology Press, s.l.
15. Zhang Y et al (2012) A survey of cyber crimes. Secur Commun Netw 422–437
16. Smadi SM (2017) Detection of online phishing email using dynamic evolving neural network based on reinforcement learning. s.l.:s.n
17. Jain AK, Gupta BB (2016) A novel approach to protect against phishing attacks as client side using auto-updated white-list. EURASIP J Inf Secur
18. InformationWeek (2005) California enacts tough anti-phishing law. https://informationweek.com/california-enacts-tough-anti-phishing-law-/d/d-id/1036636. Accessed 15 Nov 2020
19. Qabajeh I, Thabtah F, Chiclana F (2018) A recent review of conventional versus automated cybersecurity anti-phishing techniques. Elsevier Inc., pp 44–45
20. ClickDimensions (2014) Canada's anti-spam law (CASL). http://www.clickdimensions.com/sites/default/files/PDF/WhitePaper-CASL.pdf. Accessed 20 Nov 2020
21. Chuenchujit T (2016) A taxonomy of phishing research. Illinos: s.n
22. Sheng S et al (2007) Anti-phishing phil: the design and evaluation of a game that teaches people not to fall for phish. s.l., s.n
23. Arachchilage NAG, Love S, Beznosov K (2016) Phishing threat avoidance behaviour: an empirical investigation. Elsevier 60:185–197
24. Sheng S et al (2009) An empirical analysis of phishing blacklists. In: Proceedings of sixth conference on email and anti-spam (CEAS)
25. Mohammad RM, Thabtah F, McCluskey L (2015) Tutorial and critical analysis of phishing websites methods. Comput Sci Rev (17):1–24
26. Chen J, Guo C (2006) Online detection and prevention of phishing attacks (Invited Paper). IEEE Xplore
27. Wang Y, Agrawal R, Choi B-Y (2006) Light weight anti-phishing with user whitelisting in a web browser. IEEE, pp 1–7

28. Yu WD, Nargundkar S, Tiruthani N (2009) PhishCatch—A phishing detection tool. IEEE.
29. Cook DL, Gurbani VK, Daniluk M (2008) Phishwish: a stateless phishing filter using minimal rules. s.l., In: International conference on financial cryptography and data security
30. Zhang Y, Hong J, Cranor L (2007) Cantina: a content-based approach to detecting phishing websites. ACM, pp 639–648
31. Dunlop M, Groat S, Shelly D (2010) GoldPhish: using images for content-based for phishing analysis. IEEE, pp 123–128
32. Wenyin L et al (2005) Detection of phishing webpages based on visual similarity. ACM.
33. Rosiello APE, Kirda E, Ferrandi F (2007) "A layout-similarity-based approach for detecting phishing pages." In: 2007 Third International Conference on Security and Privacy in Communications Networks and the Workshops-SecureComm 2007, pp 454–463. https://doi.org/10.1109/SECCOM.2007.4550367
34. Geng G-G, Lee X-D, Wang W, Tseng S-S (2013) Favicon—A clue to phishing sites detection. IEEE, pp 1–10
35. Alghoul A et al (2018) Email classification using artificial neural network. Int J Acad Dev 2:8–14
36. Somesha M, Pais AR, Rao RS, Rathour VS (2020) Efficient deep learning techniques for the detection of phishing websites
37. Khonji M, Iraqi Y (2011) A brief description of 47 phishing classification features
38. Azad B (2011) Identifying phishing attacks, pp 1–5
39. Dewan P, Kashyap A, Kumaraguru P (2014) Analyzing social and stylometric features to identify spear phishing emails. IEEE, pp 1–13
40. Jameel NGM, George LE (2013) Detection of phishing emails using feed forward nueral network. Int J Comput Appl 77
41. ALmomani A et al (2013) Phishing dynamic evolving neural fuzzy framework for online detection "Zero-day" phishing email. Indian J Sci Technol 6(1)
42. Kathirvalavakumar T, Kavitha K, Palaniappan R (2015) Efficient harmful email identification using neural network. Br J Math Comput Sci 7:58–67
43. Duman S et al (2016) EmailProfiler: spearphishing filtering with header and stylometric features of emails. IEEE
44. Smadi S et al (2015) Detection of phishing emails using data mining algorithms. IEEE, pp 1–8
45. KDD (2020) Overview of the KDD process. http://www2.cs.uregina.ca/~dbd/cs831/notes/kdd/1_kdd.html. Accessed 2020
46. Jain A et al (2020) Overview and importance of data quality for machine learning task. ACM, USA
47. Zhang S, Zhang C, Yang Q (2003) Data preparation for data mining. Taylor & Francis, s.l.
48. BREIMAN L (2001) Random Forest. In: Schapire RE (ed) Machine learning. s.l.:Kluwer Academic Publishers, pp 45, 5–32
49. Yasin A, Abuhasan A (2016) An intelligent classification model for phishing email detection. Int J Netw Secur Appl (IJNSA) 55–70
50. Maalouf M (2011) Logistic regression in data analysis: an overview. Int J Data Anal Tech Strateg 3:281–299
51. Harrison O (2018) Machine learning basics with the K-Nearest neighbors algorithm. https://towardsdatascience.com/machine-learning-basics-with-the-k-nearest-neighbors-algorithm-6a6e71d01761. Accessed Dec 2020
52. Altaher A (2017) Phishing websites classification using hybrid SVM and KNN approach. Int J Adv Comput Sci Appl
53. Kurtanović Z, Maalej W (2017) Automatically classifying functional and non-functional requirements using supervised machine learning. IEEE.
54. Clement J (2020) Number of e-mails per day worldwide 2017–2024. Available at: https://www.statista.com/statistics/456500/daily-number-of-e-mails-worldwide/. Accessed Jan 2021
55. Retruster (2019) Phishing and email fraud statistics 2019. https://retruster.com/blog/2019-phishing-and-email-fraud-statistics.html. Accessed 2020

The Impact of GDPR Regulations on Cyber Security Effectiveness Whilst Working Remotely

Amin Dangheralou and Hamid Jahankhani

Abstract In light of the industry 4 evolution, the concerns regarding the data privacy seemed to rise due to the vast platforms and the adverse impacts that its data protection policies can have on people using it. In 2018, the European Union imposed a regulation called the general data privacy regulation (GDPR). The regulations aimed towards the protection of digital data that was transferred outside of Europe. The regulations focused on the safety of the digital data of users, the security of their data in various digital sites that they are continually sharing. They also aimed to ensure that the privacy laws concerning digital data were vigilant in modifying their rules as the digital technological changes occur in the digital world. The final aim of the GDPR is to certify that regulated legislation is unified in every region of Europe and are being followed and implemented by the digital platforms operating in Europe. As the Pandemic took hold from late 2019, there are questions arising in Europe as how to evaluate the effect of pandemic and the GDPR impact it will have as COVID 19 has forced many to go into a lockdown and as a result work remotely. The research aims to evaluate and measure the impact of GDPR on cyber security effectiveness whilst working remotely.

Keywords Pandemic · GDPR · Working remotely · Cyber security · Data protection · Privacy · Regulations

1 Introduction

The need for the GDPR was necessary to ensure that the businesses and citizens were aware of the protection of their digital data which can be stolen easily. Thus, the act was made public and provided in-depth guidance to the citizens and businesses such that they are aware of the cyber security that they need to implement into their places [1]. The law made sure that businesses complied with the regulation, especially the companies which are selling personal data of the citizens of Europe [2]. The locations

A. Dangheralou · H. Jahankhani (✉)
Northumbria University, London, UK
e-mail: Hamid.jahankhani@northumbria.ac.uk

© The Author(s), under exclusive license to Springer Nature Switzerland AG 2021
R. Montasari and H. Jahankhani (eds.), *Artificial Intelligence in Cyber Security: Impact and Implications*, Advanced Sciences and Technologies for Security Applications, https://doi.org/10.1007/978-3-030-88040-8_10

are strictly regulated to be compliant with the GDPR. The GDPR guidelines for the businesses include that the citizens will have the right to access to the data that the businesses are using. The business is obligated to provide a detailed overview of the usage of the data of the citizens. The companies are also regulated to delete the data of the citizens who have withdrawn their consent regarding the data [3]. The citizens were also empowered by the act to remove their data from one service provider to another. The further guidelines of the regulation included the right to object, right to be informed, right to have the information corrected, right to object.

These guidelines enabled the citizens to be more aware of the happening to their data and regulated cybersecurity in Europe. The legislation is also implemented to all sorts of businesses that are operating in Europe, especially the ones working remotely needs to be compliant with the GDPR requirement. A survey concluded that only 20% of the businesses were compliant with the regulation, whilst 53% of these businesses were in the implementation phase [3]. The regulation needs to be implemented in all sorts of businesses that are working on storing and using the personal data of citizens. However, the liability lies more on the business that is working remotely as they are using the personal data of citizens without any physical existence [4]. The legislation also regulates these businesses to provide full closure to their customers regarding the guidelines of the regulations.

The research is aiming towards describing the significant guidelines of GDPR and its overall impacts on the businesses that are working remotely. The regulation is essential to the companies and more importantly, to the citizens who need to be aware of their rights regarding cybersecurity. The current study shall help to provide the necessary awareness that citizens of Europe need to have regarding their personal data. The Guidelines set by GDPR has provided in-depth understanding and empowerment to citizens. The current research will focus on analysing the effects of the regulation on cybersecurity for the businesses that re operating digitally. The study shall provide essential insights on the impact that GDPR have on the firms operating remotely and cybersecurity which are provided to the citizens regarding their rights in terms of digital information.

2   Literature Review

2.1   Digital Privacy Regulations of Europe

Modern computers, especially mobile devices, have significantly improved accessibility for consumers around the world. Consumers use banking services around the globe to manage their finances or shop online to avoid going to physical shopping; People have amassed many online credentials to improve the efficiency of many day-to-day tasks [5]. Consumers do not use a single device but often use different personal devices from different companies. The goal of a modern computer is to increase value and simplify data exchange and business. To complicate matters further, these options are not limited to the organisation and its authorised users, but also hackers/cybercriminals. Unnecessary and mixed equipment of individuals and companies as well as a growing number of cloud data companies have significantly expanded the scope of the attacks, increased the complexity of the agency's protection and reduced complex attacks [6]. When organisations seek to create resistance for unauthorised users by using advanced technologies and hiring qualified cybersecurity professionals, the European Union (EU) has issued a regulation aimed at harmonising protection laws data across Europe to protect, improve and modify data privacy for all EU citizens, how local authorities ensure the privacy of the data. Although the EU has adopted a data protection law since the 1980s, this is the first article directly related to it. These are the first European laws in the world to protect the privacy and security of data.

But before the adoption of the GDPR, there were no harmonised rules on the management of personal data. Indeed, even the Data Protection Directive adopted by the EU Member States in 1995 is interpreted inconsistently and leads to different approaches. That is why the European Commission in Brussels in January 2012 proposed a revision of the 1995 Privacy Policy from 1995 to "adapt Europe to the digital age" [7]. Therefore, the Commission follows the rules (including existing laws) rather than directives (laws that can be interpreted separately by the Member States). The EU believes that with this new rule, it can eliminate fragmentation and create a single window for data protection in Europe. The main goal of reform is to protect people's right to personal data better. The GDPR defines personal information as an individual or any information relating to the data subject that may be used to identify the individual directly or indirectly.

As all organisations processing personal data related to EU citizens must comply, the GDPR is the world's first data protection law. Also, everything must be documented, make a risk assessment and take appropriate technical and organisational measures to deal with residual risk. Due to the required level of accurate documentation, institutions are unlikely to comply with GDPR obligations and use spreadsheets and word processing documents to demonstrate compliance [8]. Verification must be independent of consistency, and therefore, correct and complete documentation is necessary to reduce audit costs and the results of audits, follow-up and research. When assessing GDPR risks and identifying appropriate technical and organisational measures, an organisation must understand the risks from a business perspective.

Organisational measures include policies and procedures for screening new hires and handling or accessing data from third parties internally, monitoring the entry and modification of data, and coordinating the input file and the provision of information, education and training of staff on SDLC's confidentiality and procedures includes an information security assessment. Third-party management, the agility of the company, etc. [9]. Record the results of routine testing of technical and organisational measures to ensure continuous planning and effective implementation. The test can be performed manually or by automatic continuous monitoring. Monitor the overall status of the risk profile in GDPR, and for most organisations, the GDPR risk is not the same. As the amount of data discussed changes, the Agency's products and processes change, the geographical presence and dependence of third parties and the Agency's risk profile according to the GDPR. Following the risk assessment and monitoring tests, the technical and organisational deficiencies that need to be corrected to ensure GDPR compliance will be reported. By combining all the information on GDPR compliance in one platform, the user can not only demonstrate GDPR, but they can also use it when risks increase, organisational changes need to be addressed. Problems arise that need to be actively solved. Knowing that security incidents are monitored and resolved must actively monitor security incidents and report them to authorities and customers who may be affected [10].

The UK believes that identification is the most important threat. In 2015, 63% of confirmed crimes led to credible leaks, theft and weak passwords were used in 81% of cybercrime crimes [11]. This highlights the need for stronger authentication equipment that is suitable for end-users but remains safe and compliant with company guidelines and regulations. Establishing a robust personality and access control (IAM) is important to reduce authentication risk as hackers can use this risk to infiltrate and steal personal information. IAM solutions can help organisations address three major challenges in the protection of personal and confidential information. First, organisations must provide easy and secure access so users can find the information they need (whether the application is indoors or in the cloud) and persuade people to do so. Second, the organisation must ensure that customers have the necessary access rights to carry out their work. This includes requesting, viewing, providing and revoking user access, use automated processes that enable business owners to make access decisions. Finally, a review of the regulatory framework is necessary to demonstrate to regulators how institutions comply with the GDPR [12]. By combining reporting and identification management with the policies of credit rating agencies, reports can be organised more easily and efficiently to demonstrate continued consistency.

All three elements provide visibility and control so that the business can keep the pace with the changing environment. Damaged invoices, stolen judgments or mismanagement of settings can be considered defects of the GDPR. Companies must demonstrate that they use a preventative approach to managing access to personal data. The data is also divided into groups that are particularly interested in DCMS. These groups were chosen to understand better how the impact of GDPR changes under different circumstances [13]. These include individuals who have experienced cybersecurity incidents, passed privacy impact assessments (DPIA) or personal processing data, as well as large companies, large supply chain companies that are complex and interconnected, and providers of managed services (MSP), local governments (LA) and non-profit organisations providing basic services, small and medium-sized enterprises and various industries. Studies show that the effects of GDPR depend on certain characteristics of the agency.

DPIA host organisations, personal data processing organisations that influence cyber security incidents are more likely to improve their cyber security measures over the last three years. This is important to keep in mind, but there is some overlap between these three groups. This shows that the GDPR has succeeded in persuading regulators to improve network risk management. Experiential events also seem to force organisations to take action, suggesting that when organisations realise the damage caused by crime, they are more likely to make amends. Gathering this information for the organisation of the event helps them to encourage them to do something in the future without the need for direct participation in the event [14]. In the current research, the researcher also found that improvements were not achieved equality in all aspects of network security. Although key research has shown that most organisations have improved network risk management over the past three years, governance, risk management, data security and reporting have improved system security where there are changes in procurement and participation in the supply chain. Institutions are more likely to make changes to data protection than other aspects of network security. This demonstrates that organisations can benefit from an agile approach and underline the importance of detecting cyber security, responsiveness, recovery and prevention. The changes introduced by the GDPR occur in the vast majority of institutions, including 84% of all respondents surveyed by the board and staff [15].

Concerns about resilience are associated with the ongoing cost of maintaining staff consistency and raising awareness of the GDPR as an ongoing issue. It is too early to determine whether these changes have resulted in long-term behavioural changes or cultural changes towards stronger habits, and this is a potential area for future research. In most cases, when organisations have not changed their cyber security practices in the last three years, it is because they consider the current measures to be adequate (61% of all respondents in both surveys) [16]. Respondents believe that their organisations are capable of managing risk, defending against attacks, identifying threats and minimising the impact of incidents because they have strict policies and procedures and employees have appropriate expertise in cyber security. But with some organisations, that trust can be lost, and they will continue to be assisted in assessing the risk profile and the importance of the measures taken.

Other studies from the International Network Risk Survey from 2019 showed that the proportion of companies that assess network risk among the five main companies in their organisation would increase from 62% in 2017 to 79% in 2019 [17]. This does not necessarily mean better network risk management and network risk. The survey found that only 38% of respondents see GDPR as a force in any field to increase investment in online risk and that companies' have confidence in their potential for risk management has diminished. Other recent studies suggest that changes in network risk management may be partly related to the implementation of the GDPR. For example, cyber security surveys found that 38% of UK companies and 42% of UK charities changed their cyber security policies and procedures [18]. In some cases, the GDPR is responsible for the changes that have been made.

However, there is no data in the current literature to assess the impact of GDPR in the UK. The study also highlights other factors that influence network security management, investment and behaviour change. This can be seen from the International Online Survey, which identified cyber incidents as one of the main drivers of increased investment in cyber risk management. This study suggests that organisations may be a passive approach to cyber security and that stricter rules on cyber security are less regulatory, such as regulations such as GDPR. Furnell and Shah [19] study in ten countries in Europe, America and Asia showed that GDPR drives internal processes above expectations. It turned out that 91% of policymakers reported better processing and management of personal information [20]. Many areas have benefited from this, he says, including the transformation of information technology, cyber security and organisational change. However, a UK cyber security study found that (as last year), although most organisations have technical oversight such as security settings, firewalls and malware protection, they are unlikely to have a public cyber security policy, especially concerning portable materials or projects.

Research shows that people are more aware of information security. Following the introduction of the GDPR, the number of reports from all ICO processors has quadrupled. This suggests that more emphasis has been placed on detecting and reporting cyber security incidents related to personal data, which is likely due to new reporting requirements for GDPR breaches. However, the ICO noted that no more than 82% of reports of breaches of personal data required action by organisations since the GDPR came into force and initially pointed out the problem of reporting in September 2018 [21]. This was the International Network Risk Assessment 2019, and there may be some differences in accounting and therefore managing network risk. The results of this study show that organisations place more emphasis on technology and prevention than on prioritising the time, resources and actions needed to improve network security. Therefore, organisations believe that they can eliminate or reduce network risk primarily through technology rather than through planning, transport and response measures.

The size of the agency may also affect the impact of the GDPR. Research on GDPR as an indicator of risk management surveyed 1,300 executives and found that respondents from large organisations were more likely to report higher levels of GDPR [22]. This is due in part to increased capital to invest in compliance and administrative infrastructure to support regulatory action. Also, the report highlights the fact that many large companies do extensive business in the United Kingdom. In these countries, people have developed and implemented data protection measures and violate reporting rules. As a result, they are more likely to have a strong infrastructure and are easier to adapt to GDPR requirements. The survey found that 49% of large companies have changed their cyber security controls to comply with the GDPR, while only 28% of medium-sized companies have changed [23]. When asked about the main reasons for buying computer insurance, the number of large companies that mentioned explanations or uncertainty in rules (such as GDPR) was seven times greater than medium-sized companies.

It also urges executives as a whole to believe that there is no need to invest in cyber security and to believe that without weaknesses (as far as is known) this will not happen. Also, a survey by Grant Thornton in 2019 showed that more than 60% of companies in the survey indicated that no board member has special responsibility for network security. Approximately the same number of companies do not regularly conduct formal audits of cyber security risks and controls. This is also demonstrated in the 2019 International Network Risk Perception Survey, which found that while network risk is high on the agency's priority list, risk management has identified information technology as the main owner [23]. This view shows that the continuing lack of understanding of network risk is primarily a technical issue and not a significant business risk that requires the use of strategies to manage the company's strategic risk.

About one-third of respondents (34%) said that their organisation had received specialised training in cyber security, such as password protection, access control, patching and phishing prevention. This does not apply to courses that only cover network security aspects (such as GDPR training). Organisations that have experienced cyber security incidents (57%), passed DPIA (45%) or processed personal information (38%) are more likely to receive specialised cyber security training than typical respondents [24]. As shown by [25], when organisations provide specific cyber security training, this training is usually delivered over the last three years (83%). In the last three years, this type of training has also been organised or intensified by organisations that process personal data (86%) and organisations that have completed AIPD (85%) as the average respondent (83%). This is more common among healthcare professionals, 98% compared to 83% of all employees in the survey. From an industrial point of view, manufacturing organisations (61%), information and telecommunications (51%) and finance and insurance (48%) offer targeted network security rather than construction organisations [26]. As mentioned above, 46% of board members allowed the development of employees in network security. Almost half of the board members also indicated that the association's investment (53%) and coverage (50%) in cyber security education had increased in the last three years [24].

2.2 Impact of General Digital Protection Regulation on Cyber Security

Cyber security threats are a global problem, and this fact was recognised relatively early in the EU and its various institutions. Furthermore, it has become clear that the problem can only be tackled with comprehensive responses that require international trade, harmonised legislation and public–private partnerships. However, network security issues are very complex and sometimes difficult to implement an integrated approach. To address this, the European Commission published a message in 2001 on the European transition to information society [6]. These communications refer to many existing methods and provide additional measures to protect the information and communication infrastructure. It requires the adoption of a comprehensive policy, a common definition of cybercrime, closer communication with many stakeholders and increased funding for research and development to address these threats.

In developing its cyber security policy in 2013, the European Union stated its position on co-operation and exchange of views on cyber security issues. This will be in the form of a comprehensive platform called the Networking and Information Solutions (NIS) platform to promote the use of secure IT solutions and improve the cyber security effectiveness of IT products in use. To date, many European Member States have used the national transposition directive NIS and therefore, in 2017, the Commission approved a network security package. For example, there are proposals to build stronger and better cyber security institutions in the EU and EU Member

States [8]. Also, a European cyber security certification system should be introduced that maximises the concept of response to incidents and requires legislation and a framework to combat tax evasion and counterfeiting, as well as money, to reduce cybercrime. The Committee also called for increased international cooperation on cyber security (including the EU and NATO), the development of cyber security skills for civilians and military personnel, and the establishment of a training and education platform.

Based on these guidelines, ENISA, established in 2004, has been recognised as a leading body of the European Union Network Security Agency, mainly through information and advice. Furthermore, the bill strengthens and regulates the security certification system previously proposed by the EU for ICT products. Together with the European Commission and ENISA, the Criminal Investigation Network (T-CY) presents the Budapest Convention on cybercrime with the EU, which is a member of the European Union. And there is a big difference between management methods and priorities [11]. The EU has already acknowledged these difficulties by taking steps to address cyber threats. Among other things, they will focus on promoting democratic sustainability, in particular on measures to strengthen electoral structures and campaign information. Furthermore, it needs further guidance on the application of EU data protection law and guidance on legislation to facilitate coordination between the EU Member States on cyber security issues a network of national contacts for collaboration, research and innovation.

For example, Luxembourg's network security policy poses many important challenges for the country. It also specifies the action plan of the responsible authority and the estimated delivery schedule. Furthermore, the policy recommends and requires information, training and awareness of online risk. The update from 2018 underlined this once again and called for measures to increase public confidence in the digital environment and better protect digital infrastructure. Among them, the Luxembourg 2018 plan is one of the few new methods compared to other EU member states [14]. Harmful behaviour, cybercrime and protection of digital life, privacy and personal information of British citizens. Furthermore, UK approaches to addressing and mitigating cyber threats include international management support and the transfer of security education, services and know-how and support for products, including public education. To achieve the above personal goals, however, the UK does not see precise strategic units such as Luxembourg.

As mentioned, facts have shown that although many countries have such a realistic understanding of the meaning and scope of cyber security and cybercrime, it is difficult to find out. As a result, cyber security plans in these countries vary greatly in detail and interaction. This problem can take time, wider pan-European communication and better coordination. Most EU member states have established organisations that deal with cyber security, such as the BSI in Germany (Federal Agency for Information Security) [18]. The Agency's task is to examine the current security risks in information technology and to prepare an annual report on the state of security aspects in Germany. It also acts as a cyber-defence and security incident reporting service. BBK (Office of Civil Defence and Emergency Assistance), along with another agency, provides BSI with an Internet platform to protect critical infrastructure.

In addition to organisations such as BSI, there are many EU countries/regions with national expert groups dealing with security issues. These specialised teams are organised by the IT Emergency Response Team (CERT), sometimes referred to as the Emergency Response Team or Security Event Response Team (CSIRT). They are linked globally and across the European Union to provide safety warnings and troubleshooting, especially for product safety teams involved in government, business and academia. There are many unresolved issues in addressing national security at the national and institutional levels within the framework of harmonised policies and policy-making decisions. For example, there may be problems related to the ambiguity of regional jurisdictional conflicts and the institutional role of the German ISB, which seeks both offensive and defensive objectives. Besides, the German government established other institutions in 2017 and 2018 and is now responsible for the development and defence of cyber security plans and measures [20]. For example, in August 2017, the German government established the German Science Center (Zitis) to develop new information and law enforcement tools. Furthermore, in August 2018, the establishment of a new cyber security institution was announced for research into cyber security and key technologies.

Although the UK was only one of many EU countries, and this is just one example, it shows how governments can effectively define areas where coordination and simplification of systems can overlap or even conflict. Therefore, the EU cyber security policy must take into account all fundamental rights, including all measures [19]. In 2015, the next SEPD employee, Giovanni Buttarelli, emphasised this need in his subsequent comments on national security in 2015. Until now, the European Union has also recognised the need to strengthen the protection of personal data. This is the main reason why the European Union has started reforming its data protection system but continues to implement the new privacy and telecommunications rules.

2.3   Impact of GDPR on Businesses and Citizens Whilst Working Remotely

In recent years, when advances in technology can help employees communicate and improve productivity outside the office, teleworking has become an increasingly popular choice among organisations. With the COVID-19 pandemic, this technology is more important than ever. Social distance forces many of the citizens to stay at home for the foreseeable future, which means that teleworking is no longer an option for many, but a necessity. Without the security features that come with a desktop computer (such as whitelisted IP addresses), it is vulnerable to several exposures. Therefore, information security should be a priority at this time [26]. When faced with many other issues, the user needs to make sure that the data leak is last. Every time an organisation creates a new way to access its data, it is at greater risk. Remote processing increases this risk as it can be difficult for employees and organisations to learn about data breaches or even more difficult to identify the source of the data.

Organisations need to patch vulnerabilities in their networks and preserve physical data. When this happens, most remote workers need to transfer data or computers that can access the data to a public place. This increases the risk of data loss, and many crimes were committed because of documents left in the train, USB sticks falling from someone's pockets or a stolen laptop. While it is difficult to prevent the misuse of personal information there is nothing that companies can do to establish strict privacy data; there are still opportunities to reduce losses after data has been compromised [27]. The introduction of strict access rights means that when criminals seize an employee's laptop or other work equipment, they only see a fraction of the company's data.

To prevent the misuse of laptops and devices, companies can generally use software to monitor how employees (or criminals) use the device. Many programs can record keystrokes or monitor the movements of mice, but this can cause problems following the GDPR. Teleworkers are more likely to stick around for a long time and use their equipment for personal and professional purposes, so they cannot distinguish between looking at their work and their private lives. Therefore, it is not possible to monitor equipment unless the privacy of employees is impaired [7]. In certain cases, the responsible Director must carry out an impact assessment on data protection. For example, a company's IT department is used to assess risks associated that could cause a company financial or reputational damage, therefore, it is necessary to implement existing technologies and measures. If the processing is considered to involve a significant risk to the rights and freedoms of individuals, an additional assessment of the impact on privacy must be made.

In addition to the technical and organisational measures required for preventive and indirect data protection, data controllers and processors must report data breaches in certain circumstances and within a specified time. However, in the case of high risk, the stakeholder must also be notified directly without undue delay, unless special technical and organisational measures have been taken which make it difficult to understand and do not have access to, as well as encryption. Also, if the driver has taken the following measures to avoid this high risk, or if the workload associated with the message is too high, the message could be discarded [10]. In the latter case, however, the responsible party may need to adopt public relations or other similar measures. The Data Protection Supervisor has strengthened his enforcement powers with a new legal framework (including broader fines). Therefore, it is recommended that all data managers have an effective data protection management program in their organisation.

Network security is not only a problem for the police and national security but also the EU. Rather, it is a global problem that has led public and private entities to consider the best risk mitigation programs. Among them, government policies and strategies related to cyber security have attracted the attention of European citizens, as cyber security incidents often involve the loss, damage or unauthorised disclosure of the personal information [13]. In the field of cyber security, the European Network and Information Security Agency (ENISA) has developed a classification theory to classify different types of threats and individual threats in different details. The purpose of this classification is to establish housing benchmarks.

The World Economic Conference (WEF), a non-profit Swiss organisation, seeks to bring together business leaders, policymakers, academics and others to discuss global, regional and business agendas. From their point of view, an event can cover a wide range of areas, such as data breaches and encryption, or identity theft, and there can be for a variety of reasons. The causes and consequences are numerous, generally unexpected. Besides, in its International Risk Report from 2017, the World Economic Council has identified 12 key issues that may play a role in future cyber security [17]. The owners of the devices rarely notice a malware infection because their computers can continue to function except for secrecy and increased use of Internet bandwidth. As a result, cyber security incidents affect many different technical areas of civil society and government, and it is, therefore, important to take appropriate measures to ensure the availability, integrity and confidentiality of these technologies. This includes individuals with personal information collected and processed digitally, who may be at risk.

Although private individuals can carry out cyber-attacks to obtain financial or social incentives, government activities generally extend to more diverse law enforcement agencies, including law enforcement agencies. Laws (LEA) that organise online worlds for the benefit of criminal investigations or prevention and national security. Targeting can also vary, and attacks on critical infrastructure are considered the most pressing problem in every country in the world, followed by attacks on government agencies themselves, such as attacks on government agencies [21]. The use of so-called security monitoring technology (SOST) plays an important role when the government pays special attention to potential computer security personnel. Furthermore, the use of multiple criteria in many EU countries is allowed to varying degrees, which is often criticised by the media and human rights advocates. The citizen-cantered approach has provoked controversy over the dangers of violence and prejudice, which are often accompanied by a significant lack of transparency.

The media and human rights organisations also criticise government agencies for using so-called zero-day attacks to gain influence over domestic and foreign intelligence services. As the entire information technology environment becomes increasingly uncertain, these methods receive considerable attention, but at the same time provide vulnerabilities that can be detected and potentially exploited, not only by legitimate national security organisations but also by bad people [15]. In this case, a general discussion about legal access for police and intelligence is also important. For a long time, many of these organisations have been requesting access to encrypted devices with external features. As a result, a company's legal obligations to manage such access in the future may affect any software, even hardware.

In November 2016, the Council of the European Union proposed to launch a debate on these issues under the leadership of the European Commission. The issue of encryption was then discussed at a meeting of the Council on 8–9 December 2016 [25]. At the meeting, the ministers agreed that caution should be exercised when addressing privacy risks and network security. In December 2016, ENISA also published a conclusion on the encryption document and concluded that weakening the encryption to ensure legitimate eavesdropping is not the best way. ENISA clearly warns of unforeseen consequences, such as the weakening of digital signatures, and recommends further analysis of benefits and risks and a thorough study of options before taking possible action in legislation.

Similarly, in 2014, the European Group on Ethics for Science and New Technologies (EGE) published opinions on safety and surveillance technologies, highlighting the risks associated with these technologies. Li et al. [28] emphasised that although non-residents can create problems, it should be borne in mind that there are risks when using aggressive surveillance technology on land. Therefore, the principles and values of Europe, as well as democracy, must be taken seriously.

Thus, especially concerning national security [27], finally returned to the issuance of restrictions on the necessity and efficiency of measures and objectives that must be allowed for national surveillance. But this is not just a decoding question, but a question of government action, including SOST. In particular, as law enforcement agencies increasingly use high-tech diagnostic tools, people are becoming increasingly concerned that citizens are unlikely to be able to protect themselves against perceived security threats or abuse based on the algorithm. This applies not only to specific prevention or investigations into the crime of law enforcement agencies but also to the interests of national security.

3   Research Methodology

This section aims to elaborate on the methodology, including the sample of the research design and the techniques utilised for conducting the research combined with survey questionnaires which is also used as a means of conducting the survey for this research. "Statistical Package for Social Sciences (SPSS)" is used in this research for the collection of data and further analysing the results. Leading to discussion and conclusion of the focused area of the research.

According to the studies conducted by [29], Research methodology refers to the preferred method type particularly utilised by the researcher to gain true and accurate data. In addition [30], states that the guidelines of the research also directs the path of the researcher in the most appropriate direction for conducting the best research. Meanwhile, Team [31] is of the opinion that there are different type of research methods utilised by researchers that usually dependent on the requirements of the research such as quantity, qualifications and the diversified methods. [32], argues that depending on the research method it is seen as a set of actions taken to perform an analysis. There are different types of research methods, including computational methods, behavioural patterns and mixed methods [33]. However, in the present study, the researcher used a more general approach. The method of quantitative approach also referred to as the method that allows the quantification and the test result.

On the other hand [34], states that the qualitative approach focuses on the ideas and opinions of the respondents. It also provides an overview of all qualitative information that may be preliminary research and observation. Meanwhile [35], states that the quantitative method, is a complex method because it allows the researchers to verify the results and confirm the hypothesis. However, in view of the focused area of the research which is to examine the impact of the introduction of the GDPR Regulations on the effectiveness of cyber security effectiveness while working remotely. The researcher recommends and utilises a comprehensive approach through research.

3.1 Data Collection

Findings from the study conducted by [36] regarding the utilisation of the most appropriate research methodology regarding study based on quantitative research methods states that the collection of data is one of the most essential sections of a research study. Moreover [34], states that due to the fact that the collection of authentic and accurate data is the foundation of the entire study and is based on the authenticity of information. Furthermore [37], reveals that the research study will consist of two types of baseline information used for their collections. For the collection of secondary information, the second most important items such as books, peer journals, articles, preliminary research, the website of various libraries and much more is utilised for obtaining the data and information concerning the general overview of the impact of the introduction of the GDPR Regulations on the effectiveness of cyber security effectiveness while working remotely. Alternatively, the first data is collected from the underlying cause through research. However, once the data has been compiled, the information and data collected will be researched and further analysed in order to make the most relevant and relevant research possible.

- Sampling Size

  Gligorijevic et al. [38] in their studies found that the sample size of the research study reflects the representative units selected from a particular group or population to carry out the investigation to satisfy the aims and objectives of the research. Therefore, with due regard to this objective, in order to process the factually obtained data by minimizing the possibility of errors through a good representation of the population, a sample size of 30 people was selected. The snowball sampling technique was also used in this research and participants are from both genders. In addition [39], states that Non-probability sampling is the snowball sampling method and participants in this sample have study-specific characteristics.

  This type of sampling is useful for this research as it can support the aim and provide a detailed insight into the impact of the introduction of the GDPR Regulations on the effectiveness of cyber security effectiveness while working remotely. As well as, they know the sample size is thirty executive level employees who are asked to answer questions about their business and the impact of the introduction of the GDPR Regulations on the effectiveness of cyber security effectiveness while working remotely. The sample size selected is non-probability sampling and therefore does not require the population to belong to a particular age range or socioeconomic status. These employees are contacted through the LinkedIn platform where it is easier for such researchers to gather their data.

- Population

  As demonstrated in the studies conducted by [40] reveal that the research population represents the entirety of the participants involved in the research study. Hence, while maintaining this, the research study aims to encompass the population comprising senior managers, and executives from different organizations.

  Limitations.

  This study highlights the preservation of the confidentiality of information provided by representatives of groups and populations who worked from home during the Covid-19 situation. The task force should focus on the main factors influencing the introduction of the GDPR Regulation on the effectiveness of cybersecurity when working remotely with the provision of only their personal information. The name of people is not mentioned to confirm their trust [41]. Representative motivation and discussion aim to obtain relevant feedback as required by the study [42].

- Data Measurement and Questionnaire Design

  According to a study by [43], the search was based on a questionnaire based on queries designed to address previous research, research, and literature related to the same subject [44]. Thus, after reviewing past studies and guidelines, as well as the requirements, and objectives of the existing study, concise, relevant, explanatory and closed questions are asked [44]. Simple English queries are used so that respondents can easily understand the questions and provide reliable, and concise answers that will ultimately facilitate the process of finding and analyzing that data.

  The content of the question will be divided into two parts. The first part focuses on the number of respondents who provide user information, and the second group asks how their information relates to the impact of the GDPR on remote service security. Meanwhile [45], have argued that the Likert scale, when compared to the scale of 1–5, is ideal for measuring data.

- Ethical Consideration

  In this study, all research was conducted accordingly on the selected topic. References in this study are quoted correctly without infringing copyright. The questionnaire was provided to the participants with the prior acceptance of consent of the respondents. In addition, the nature of the respondents and the personal data is kept confidential. According to [29, 30], research from the present study regularly involves coordinating people from different foundations, for example, from different societies, institutions, and orders that, for example, the moral character is extended, for example: duty, universal respect and trust.

  Team [31] states that many standards of moral research, such as copyright rules, patent strategies, creators, information sales agreements and corresponding rules for assessment companies, are designed to protect the interests of profit seekers by expanding cooperation at a comparable level. The basic stage at which many researchers receive their information and do not necessarily accept or disclose their ideas at random [32]. Therefore, it is important for an investigator to follow all moral considerations when considering this question, otherwise it does not matter.

According to studies by [33–35], it is important for professionals to understand their obligations regarding how moral research is shaped. The analyst must investigate moral behaviour and descriptions, including how it is conducted, and apply those research moralities in his research procedure, wherever important [46]. The specialist must detect the beach and the flow that affect the details of the moral reflection on the examination [47].

- Data Analysis Plan

The research conducted by [43], shows that good research information includes good data analysis after his good research collected from the question. In addition, SPSS-based software is used for statistical analysis and tests performed on the data include t-test, demographic analysis, regression analysis, data analysis, and more [22]. SPSS is also used to determine the relationship between independent and dependent variables and their impact on each other. After the data analysis, all the information obtained will be added to the research.

4   Data Analysis and Critical Discussions

4.1   Data Analysis

The total number of participants was n = 30 out of what 22 (73%) were male and 8 females (27%).

Social distance forces many of the citizens to stay at home for the foreseeable future, which means that teleworking is no longer an option for many, but a necessity.

		Frequency	Percentage	Valid Percentage	Cumulative Percentage
Valid	Strongly disagree	7	23.3	23.3	23.3
	Disagree	3	10.0	10.0	33.3
	Neutral	14	46.7	46.7	80.0
	Agree	4	13.3	13.3	93.3
	Strongly agree	2	6.7	6.7	100.0
	Total	30	100.0	100.0	

In response to the question, only 04 participants agreed whereas 02 strongly agreed to the statement; meanwhile, 03 participants disagreed, and 14 remained neutral, and 07 strongly disagreed. Hence only 20% of the participants were in agreement with the statement whereas, 33% remained in disagreement with the statement.

The GDPR caused many constraints for citizens working remotely.

		Frequency	Percentage	Valid Percentage	Cumulative Percentage
Valid	Strongly disagree	2	6.7	6.7	6.7
	Disagree	4	13.3	13.3	20.0
	Neutral	9	30.0	30.0	50.0
	Agree	9	30.0	30.0	80.0
	Strongly agree	6	20.0	20.0	100.0
	Total	30	100.0	100.0	

In response to question, 09 participants agreed whereas 06 strongly agreed to the statement; meanwhile, 04 participants disagreed, and 9 remained neutral, and 02 strongly disagreed. Hence almost 50% of the participants were in agreement with the statement.

Without the security features that come with a desktop computer (such as whitelisted IP addresses), it is vulnerable to several exposures.

		Frequency	Percentage	Valid Percentage	Cumulative Percentage
Valid	Disagree	1	3.3	3.3	3.3
	Neutral	4	13.3	13.3	16.7
	Agree	16	53.3	53.3	70.0
	Strongly agree	9	30.0	30.0	100.0
	Total	30	100.0	100.0	

In response to the question, 16 participants agreed whereas 09 strongly agreed to the statement; meanwhile, 01 participants disagreed, and 04 remained neutral, and none strongly disagreed. Hence almost 66% of the participants were in agreement with the statement.

The GDPR focus on the safety of the digital data of users, the security of their data in various digital sites that they continually share.

		Frequency	Percentage	Valid Percentage	Cumulative percentage
Valid	Disagree	2	6.7	6.7	6.7
	Neutral	12	40.0	40.0	46.7
	Agree	13	43.3	43.3	90.0
	Strongly agree	3	10.0	10.0	100.0
	Total	30	100.0	100.0	

In response to the question, 13 participants agreed whereas 03 strongly agreed to the statement; meanwhile, 02 participants disagreed, and 12 remained neutral, and none strongly disagreed. Hence almost 83% of the participants were in agreement with the statement.

GDPR ensure that the privacy laws concerning digital data are vigilant in modifying their rules as the digital technological changes occur in the digital world.

		Frequency	Percentage	Valid Percentage	Cumulative percentage
Valid	Disagree	1	3.3	3.3	3.3
	Neutral	10	33.3	33.3	36.7
	Agree	10	33.3	33.3	70.0
	Strongly agree	9	30.0	30.0	100.0
	Total	30	100.0	100.0	

In response to question, 10 participants agreed whereas 09 strongly agreed to the statement; meanwhile, 01 participants disagreed, and 10 remained neutral, and none strongly disagreed. Hence almost 63% of the participants were in agreement with the statement.

GDPR certifies and regulates legislation which is unified in every region of Europe and is being followed and implemented by the digital platforms operating in Europe.

		Frequency	Percentage	Valid Percentage	Cumulative percentage
Valid	Strongly disagree	2	6.7	6.7	6.7
	Disagree	6	20.0	20.0	26.7
	Neutral	13	43.3	43.3	70.0
	Agree	8	26.7	26.7	96.7
	Strongly agree	1	3.3	3.3	100.0
	Total	30	100.0	100.0	

In response to the question, 08 participants agreed whereas 01 strongly agreed to the statement; meanwhile, 06 participants disagreed, and 13 remained neutral, and 02 strongly disagreed. Hence almost 50% of the participants were in agreement with the statement.

The main goal of GDPR reform is to protect people's right to personal data better.

		Frequency	Percentage	Valid Percentage	Cumulative percentage
Valid	Disagree	2	6.7	6.7	6.7
	Neutral	10	33.3	33.3	40.0
	Agree	8	26.7	26.7	66.7
	Strongly agree	10	33.3	33.3	100.0
	Total	30	100.0	100.0	

In response to the question, 08 participants agreed whereas ten strongly agreed to the statement; meanwhile, 02 participants disagreed, and 10 remained neutral, and none strongly disagreed. Hence almost 60% of the participants were in agreement with the statement.

With the COVID-19 pandemic, technology is more critical than ever.

		Frequency	Percentage	Valid Percentage	Cumulative percentage
Valid	Strongly disagree	2	6.7	6.7	6.7
	Disagree	1	3.3	3.3	10.0
	Neutral	12	40.0	40.0	50.0
	Agree	12	40.0	40.0	90.0
	Strongly agree	3	10.0	10.0	100.0
	Total	30	100.0	100.0	

In response to the question, 12 participants agreed whereas 03 strongly agreed to the statement; meanwhile, 01 participants disagreed, and 12 remained neutral, and two strongly disagreed. Hence almost 50% of the participants were in agreement with the statement.

Reliability Statistics.

Cronbach's Alpha	N of items
0.919	14

Cronbach Alpha test is conducted to find the reliability of the survey conducted and a generally accepted rule is that the alpha value must not be less than 0.6. At the same time, the results of the test conducted depict 0.919 value which means that the survey conducted is highly reliable.

Correlations.

		IV_CyberSecurity	IV_Impact_on_Businesses	IV_Impact_on_Citizens_Working_Remotely
IV_CyberSecurity	Pearson correlation	1	−0.167	−0.121
	Sig. (2-tailed)		0.379	0.525
	N	30	30	30
IV_Impact_on_Businesses	Pearson correlation	−0.167	1	−0.089
	Sig. (2-tailed)	0.379		0.641
	N	30	30	30
IV_Impact_on_Citizens_Working_Remotely	Pearson correlation	−0.121	−0.089	1
	Sig. (2-tailed)	0.525	0.641	
	N	30	30	30
DV_GDPR_Regulations	Pearson correlation	0.060	−0.059	−0.099
	Sig. (2-tailed)	0.754	0.755	0.602
	N	30	30	30

The independent variable of the study is Cybersecurity, Impact on Businesses, and Impact on Citizens Working Remotely while the dependent variables were GDPR Regulations on which correlation test was run. The results from correlation clearly show that cybersecurity, impact of businesses and the impact on citizens working remotely have a negative relationship with each other. However, variables including perceived factor cybersecurity, impact of businesses and the impact on citizens working remotely all have a positive relationship with the GDPR Regulations. Before interpreting the value of correlation, it must be considered that a correlation between two variables is significant if the significant value is less than 0.05.

Interpreting the value of the correlation between cybersecurity and GDPR Regulations shows Pearson correlation value of 0.06. There is a positive relationship between the two variables, and the strength of the relationship is extreme. Similarly, the variable impact on businesses and GDPR where the correlation value is −0.059, which shows that there is a negative relationship between the two variable but with a healthy relationship. As long as the correlation is close to 0, the strength of the relationship becomes very weak, or it could be said that there may be no relationship between the variables of the study. On the other hand, interpreting the correlation values for impact on citizens working remotely (−0.099), it could be likewise stated that there is again a strong negative relationship between this variable and GDPR Regulations. Alternatively, it could be said that although the strength of the relationship is negative, it also has a stable negative relationship with two of the independent variables.

4.1.1 Regression

Model Summary

Model	R	R^2	Adjusted R Square	Std. Error of The Estimate	Change statistics		
					R^2 Change	F Change	df1
1	0.126a	0.016	−0.098	0.50340	0.016	0.140	3

According to the above table, it can be stated that most of the part of the data used in this study was valid and reliable. As the R^2 value of the survey results is 0.16, which means even thou with lower R^2 value, the regression model has statistically significant explanatory power.

Model		Sum of squares	df	Mean square	F	Sig
1	Regression	0.106	3	0.035	0.140	0.935b
	Residual	6.589	26	0.253		
	Total	6.695	29			

a.   Dependent Variable: DV_GDPR_Regulations
b.   Predictors: (Constant), IV_CyberSecurity, IV_Impact_on_Citizens_Working_Remotely, IV_Impact_on_Businesses.

The above table is about the Analysis of Variance (ANOVA) of the results obtained from the survey. As per the above table, it is found that level of significance of the study was 0.935, which means that there is a non-significant relationship between independent and dependent variables.

Coefficients.

Model		Unstandardised coefficients		Standardised coefficients	t	Sig
		B	Std. Error	Beta		
1	(Constant)	3.882	1.145		3.390	0.002
	IV_Impact_on_Citizens_Working_Remotely	−0.070	0.137	−0.100	−0.508	0.616
	IV_Impact_on_Businesses	−0.053	0.168	−0.062	−0.313	0.757
	IV_CyberSecurity	0.031	0.165	0.037	0.187	0.853

a.   Dependent Variable: DV_GDPR_Regulations

Above is the table of coefficient, which bears the significance value for dependent variables 0.616, 0.757 and 0.853, respectively. These values are representing the fact that GDPR Regulations has a substantial impact on Cyber Security, on Businesses and Citizen Working Remotely.

4.2 Critical Discussions

It was found that demonstrating GDPR and NIS legislation in the face of reduced education; youth research identifies how security can be compromised by the GDPR and other organisations, mainly the UK itself. The more people comment, the more they search for information to understand its audience, develop and selling it to the general public buy. Meanwhile, the results of this study were conducted, showing the strong correlations between cyber protections and GDPR. While number offers many benefits, it also brings with it the risks of not knowing the internet - it seems to be increasing in power, numbers and benefits. This is in line with a study conducted by [8] which states that preventive measures should be evaluated and monitored effectively and should be adequately regulated in order to be safe and reported to management and potential customers' involvement. Information based on software loyalty to paid records and secrets such as birthdays, journals and anything else used by the business to track customer information is particularly attractive to cybercriminals. Meanwhile, from the results of this study, it was found that more and more organisations are investing in cybersecurity, and many companies have reported on the progress of online security legislation since the inception of GDPR.

However [11], states that it would have been somewhat lost if the existing authorities were right (such as information security and data security and operations) if organisations had prioritised online security in the same way that like other risky businesses. Information is a valuable product when used correctly. Although the article suggests that organisations focus on technology and security, rather than prioritising the time, resources and actions needed to improve communication security. Although Yu [48] states that the GDPR is well-documented in technical issues such as machine learning, there was a limited constraint of management or incident response plans, and it was often tested and updated. Although [16] states that these options are not only for the organisation and its authorised users but for competitors/cybercriminals.

Although the GDPR has been involved in the protection of information associated with information security, results of the study found no superior force of effects on other aspects of cybersecurity in the same way, but as of the writing, the Data Protection Directive adopted by the EU Member States in 1995 means compliance and lead in different ways. Also, the 2020 Cyber Security Breach Survey suggested that there are many other things that organisations can do to improve their online security (such as auditing, cyber insurance, and consumer status risk and breach reports). The results of this study suggest that the GDPR has now made the business more responsive and also supports the general public against cybercrime. Meanwhile, as stated by [49], the GDPR-backed scripts have been successful in influencing regulation to improve risk management. In many cases, such information becomes a form of money laundering, transaction and trading on the Darknet market [22].

Besides, as noted by [10], large companies were better prepared for the launch of the GDPR due to a large number of resources and administrative controls to support compliance. As stated in the literature review by [50], Data cases often do not know this, how their information is handled or who owns it, or can make it available to them. However, it was found that, due to the need for accurate records, GDPR credit institutions may not comply and use leaflets and words trips to display them. Also, if one thinks about the ease of dealing with a brand or service that fully understands their research needs and offers a suitable solution needed, it is a 'win–win' situation for everyone. However [51], states that there was a lack of evidence as to it, how and why the GDPR impact was different for different businesses.

Meanwhile, Porcedda [46] states that above and beyond this, in some cases, organisations neglect how to manage and share information on the contrary, literature review suggests that there is a consolidation of ideas on the cybersecurity priority committee. Similarly, it was found that the fluctuating impact on business and GDPR rules set at −0.59 indicate that there is a negative correlation between the two variables, but there is a strong correlation. Unfortunately, the result of this survey found that the type of personal data that is useful to a business is also useful in the cybercriminal world and is designed for cybercrime. As suggested by [52], there are improvements in the management of cybersecurity and the status of the Committee's discussions. Inconsistent with this, the findings from this study suggest, Data protection should always be at the forefront of a business model communication model. Besides, other reports suggested that the Committee continue to ignore cybersecurity as a business risk, highlighting the many things that can be done to understand how these thoughts can change. Whereas, it was found that technology and legal imperfections needed to be corrected to ensure GDPR complied with reporting.

According to [25], Statistical numbers and its progress have led us to the middle year information cycle. Whereas, results suggest that statistical numbers quickly becomes apparent if changes have caused—long-term behavioural changes to more severe behaviours. This can strengthen and increase external sales, strengthen and capture repetitive portability in any segment, creating uncertainty about owner data in a new company. Accordingly [6], states that the organisation also noted that consumers should be assured of the right to perform their duties similarly, the literature review suggests that the support of companies should demonstrate that they use a method to prevent access to consumers' data. As such, it promotes the security and confidentiality of consumers' personal information, which is essential in the operation of hacking and reducing cybercrime. As stated by [53], the outcome of the GDPR is consistent with other organisational characteristics. The General Data Protection Regulation (GDPR) has been implemented as a way to address this problem, among others, and to ensure that data is stored effectively to protect user information. Argaw et al. [15] states that the GDPR has been successful in encouraging organisations to improve the management of supply chain management. For any cybersecurity provider who specialises in its services, the GDPR could play a key role in supporting the cybersecurity industry as a basis for respecting and protecting human privacy. Moreover [19], states that the promotion has not been achieved equally in all aspects of network security meanwhile, results show that respondents believe their organisations are capable of risk management, preventing attacks, displaying threats and minimising the risk of incidents because they have strict systems and procedures and staff with the necessary cybersecurity skills.

Respondents are of the view that remote employment has become a popular choice for organisations over the past few years, thanks to advances in technology to help employees stay connected and productive while out of the office. While it is stated under social media that many citizens stay at home for the foreseeable future, it means everyday chores are no longer common, but they should. Meanwhile Lallie et al. [26] states that these technologies are more critical than ever in understanding the COVID-19 virus. Similarly [5], states that sharing has forced many of us to stay home for the foreseeable future, meaning that working out, for the most part, is no longer an option but a necessity. As young data-savvy generations become consumers of these products, the transparency shown by companies will become more important to them, affecting their loyalty and sales decisions. As mentioned on Porcedda [46], without the security of being in the office, as if it does not have an IP address, all organisations and remote employees can easily be found in a secure security environment. While these rules need to be followed, and this is a good step in this regard, consider how the crime is happening, and how it continues to attract and attract consumers around the world. The law should not be seen as a problem.

Consistent with the statement [10, 27] state that it is challenging to prevent the misuse of public information there is nothing that companies can do to obtain reliable data There are still opportunities to control loss after data [51]. Share information with companies that are aware of the potential use of technical information in as many appropriate ways as possible, using data, and quickly, for their benefit [49]. Thus, organisations in this area have chosen, fearing where the data is growing, as well as encouraging consumers to understand why they should keep their information. Besides [53], states that knowledge, resulting in increased pressure to show compliance. The protection of information should, therefore be of paramount importance at this time. The last thing people need when they are dealing with a lot of other problems is suffering from data breaches and can cause the company financial damage or reputation [19]. Therefore, it is necessary to use existing technologies and measures. Because of this, businesses cannot rest and must act quickly to comply. The best place to start when it comes to staying safe is the General Data Protection Regulation (GDPR) as GDPR is the first data protection law [26].

Many companies continue to have problems due to the ever-increasing volume of data, which makes it increasingly difficult for entrepreneurs to see fully where the data resides and who can access it. Riddel [52] states that the GDPR is a risk, and by considering the appropriate technical and organisational measures, the organisation should understand the risk from a business perspective. It has been found that the GDPR Regulation helps organisations understand the safety risks they face and the steps they need to take to reduce them. Zerlang [22] states that in the experience, organisations have wisely developed their follow-up systems and evaluated their ability to process information so that the GDPR can be used as an opportunity to reinforce the importance of chains and present new avenues provide additional services to customers. For these organisations, the GDPR has helped to use their data to drive activity and improve trust in their businesses [15].

It was found that over the years, an increase in the number of organisations requiring ad-hoc security professionals to conduct a specialised information security impact assessment to assess the onset, nature and severity of the risks associated with the operation of consumers' data has been seen [48]. Besides, the GDPR serves as a new way of protection within the database of data centres and in cloud sites, called zero trusts [5]. GDPR is much work that only sometimes takes considerable amounts to ensure compliance. When organisations can choose the right partner and the right protection, the GDPR and increasing profitability by reducing linkages and over-activity are a good opportunity. However, many organisations, mostly small and medium-sized businesses, are changing or evaluating how they manage their data [25]. For them, the journey has now begun, but the clock is ticking to close these spaces.

5   Conclusions

On the basis of previous research, it can be concluded that the impact of the General Privacy Regulation on a digital society is indisputable and the cyber security society is no exception. However, the European cyber security community has found itself in an interesting situation looking for opportunities to meet GDPR obligations in its districts. Some people wonder if the security measures of networks and devices are in accordance with the GDPR and there are many possible problems. Network security, the field of computer security and the CSIRT team should have an extended definition of personal information that includes IP addresses and other identifiers used in network communications in accordance with the GDPR. This data can identify common users who need to maintain privacy, as well as intruders who need to be investigated and removed. When working with a network security team, GDPR issues are more obvious. CSIRT teams typically share information about current threats, incidents and countermeasures to warn others, reduce threats effectively and improve global security. There are so many applications, tools, platforms and communities that One can share almost any information One can think of. A great example is a traditional intrusion detection system or intrusion alert exchange platform, where security incident information is often shared and analyzed by multiple partners. However, some of the data that these systems exchange and analyze are personal data. Therefore, the legal aspects of these devices and processes should be taken seriously and, if possible, implemented in a constructive manner, i.e. in device development. Furthermore, the public is aware of the responsibilities of the GDPR, fear and apprehension. Different legal frameworks protect certain categories at EU level. Intellectual property rights, trade secrets and confidential information. The protection of personal data is the most likely barrier to the exchange of information. Therefore, it is important to remove these barriers so that a secure connection does not hinder the dissemination of information. To meet the needs of a new legal framework for the cybersecurity community, a data protection assessment to examine the risks of sharing cybersecurity information.

In addition, demonstrating transparency and fair consumer protection is an effective way to build trust and reputation. To remain in the EU market or remain competitive, companies around the world must intensify their efforts to manage privacy risks and protect personal information.

References

1. Ujcich BE, Bates A, Sanders WH (2018) A provenance model for the European Union general data protection regulation. In: International provenance and annotation workshop. Springer, Cham, pp 45–57
2. Labadie C, Legner C (2019) Understanding data protection regulations from a data management perspective: a capability-based approach to EU-GDPR. In: Proceedings of the 14th international conference on wirtschaftsinformatik
3. Laybats C, Davies J (2018) GDPR: implementing the regulations. Bus Inf Rev 35(2):81–83
4. Chunhsien S, Shangchien L (2018) Personal data and identifiers: some issues regarding general data protection regulations. In: International wireless internet conference. Springer, Cham, pp 163–169
5. Nielsen JC, Kautzner J, Casado-Arroyo R, Burri H, Callens S, Cowie MR, Dickstein K, Drossart I, Geneste G, Erkin Z, Hyafil F (2020) Remote monitoring of cardiac implanted electronic devices: legal requirements and ethical principles-ESC Regulatory Affairs Committee/EHRA joint task force report. EP Europace
6. Coventry L, Branley D (2018) Cybersecurity in healthcare: a narrative review of trends, threats and ways forward. Maturitas 113:48–52
7. Politou E, Alepis E, Patsakis C (2018) Forgetting personal data and revoking consent under the GDPR: challenges and proposed solutions. J Cybersecur 4(1):tyy001
8. Poulsen A, Fosch-Villaronga E, Burmeister OK (2020) Cybersecurity, value sensing robots for LGBTIQ+ elderly, and the need for revised codes of conduct. Aust J Inf Syst 24
9. Markopoulou D, Papakonstantinou V, de Hert P (2019) The new EU cybersecurity framework: the NIS directive, ENISA's role and the general data protection regulation. Comput Law Secur Rev 35(6):105336
10. Vitunskaite M, He Y, Brandstetter T, Janicke H (2019) Smart cities and cyber security: are we there yet? A comparative study on the role of standards, third party risk management and security ownership. Comput Secur 83:313–331
11. Weber RH, Studer E (2016) Cybersecurity in the internet of things: legal aspects. Comput Law Secur Rev 32(5):715–728
12. Calliess C, Baumgarten A (2020) Cybersecurity in the EU the example of the financial sector: a legal perspective. German Law J 21(6):1149–1179
13. Aliyu A, Maglaras L, He Y, Yevseyeva I, Boiten E, Cook A, Janicke H (2020) A holistic cybersecurity maturity assessment framework for higher education institutions in the United Kingdom. Appl Sci 10(10):3660
14. Jackson BW (2020) Cybersecurity, privacy, and artificial intelligence: an examination of legal issues surrounding the european union general data protection regulation and autonomous network defense. Minn J Law Sci Technol 21(1):169
15. Argaw ST, Troncoso-Pastoriza JR, Lacey D, Florin MV, Calcavecchia F, Anderson D, Burleson W, Vogel JM, O'Leary C, Eshaya-Chauvin B, Flahault A (2020) Cybersecurity of hospitals: discussing the challenges and working towards mitigating the risks. BMC Med Inform Decis Mak 20(1):1–10
16. Lykou G, Anagnostopoulou A, Gritzalis D (2019) Smart airport cybersecurity: threat mitigation and cyber resilience controls. Sensors 19(1):19
17. James L (2018) Making cyber-security a strategic business priority. Netw Secur 2018(5):6–8

18. Horák M, Stupka V, Husák M (2019) GDPR Compliance in cybersecurity software: a case study of DPIA in information sharing platform. In: Proceedings of the 14th international conference on availability, reliability and security, pp 1–8
19. Furnell S, Shah JN (2020) Home working and cyber security—An outbreak of unpreparedness? Comput Fraud Secur 2020(8):6–12
20. Gao L, Calderon TG, Tang F (2020) Public companies' cybersecurity risk disclosures. Int J Account Inf Syst 38:100468
21. Bauer TN, Truxillo DM, Jones MP, Brady G (2020) Privacy and cybersecurity challenges, opportunities, and recommendations: personnel selection in an era of online application systems and big data
22. Zerlang J (2017) GDPR: a milestone in convergence for cyber-security and compliance. Netw Secur 2017(6):8–11
23. Pesapane F, Volonté C, Codari M, Sardanelli F (2018) Artificial intelligence as a medical device in radiology: ethical and regulatory issues in Europe and the United States. Insights Imaging 9(5):745–753
24. Dwivedi YK, Hughes DL, Coombs C, Constantiou I, Duan Y, Edwards JS, Gupta B, Lal B, Misra S, Prashant P, Raman R (2020) Impact of COVID-19 pandemic on information management research and practice: transforming education, work and life. Int J Inf Manag 55:102211
25. Taeihagh A, Lim HSM (2019) Governing autonomous vehicles: emerging responses for safety, liability, privacy, cybersecurity, and industry risks. Transp Rev 39(1):103–128
26. Lallie HS, Shepherd LA, Nurse JR, Erola A, Epiphaniou G, Maple C, Bellekens X (2020) Cyber security in the age of COVID-19: a timeline and analysis of cyber-crime and cyber-attacks during the pandemic. arXiv preprint arXiv:2006.11929
27. Lim HSM, Taeihagh A (2018) Autonomous vehicles for smart and sustainable cities: an in-depth exploration of privacy and cybersecurity implications. Energies 11(5):1062
28. Li H, Yu L, He W (2019) The impact of GDPR on global technology development
29. Watson D, Millerick R (2018) GDPR and employee data protection: Cyber security data example. Cyber Secur: Peer-Rev J 2:23–30
30. Van der Ree MH, Scholte RA, Postema PG, De Groot JR (2019) Playing by the rules: impact of the new general data protection regulation on retrospective studies: a researcher's experience. Oxford University Press
31. Team IGP (2020) EU general data protection regulation (GDPR)—An implementation and compliance guide. IT Governance Ltd.
32. Sethu SG (2020) Legal protection for data security: a comparative analysis of the laws and regulations of European Union, US, India and UAE. In: 2020 11th international conference on computing, communication and networking technologies (ICCCNT). IEEE, pp 1–5
33. Ngozwana N (2018) Ethical dilemmas in qualitative research methodology: researcher's reflections. Int J Educ Methodol 4:19–28
34. Mangini V, Tal I, Moldovan A-N (2020) An empirical study on the impact of GDPR and right to be forgotten-organisations and users perspective. In: Proceedings of the 15th international conference on availability, reliability and security, pp 1–9
35. Lindgren P (2016) GDPR regulation impact on different business models and businesses. J Multi Bus Model Innov Technol 4:241–254
36. Hribar D, Dvojmoč M, Markelj B (2018) The impact of the EU general data protection regulation (GDPR) on mobile devices. J Crim Justice Secur 414–433
37. Gupta BK (2020) General data protection regulation and its impact on indian enterprises. AKGEC Int J Technol 11:28–31
38. Gligorijević M, Popović R, Maksimovića, ND, Impact analysis of the application of the GDPR regulation on the functioning of the information and communication system of the MOI of the republic of Serbia. In: The proceedings of human security and new technologies, vol 75
39. Gear C, Eppel E, Koziol-Mclain J (2018) Advancing complexity theory as a qualitative research methodology. Int J Qual Methods 17:1609406918782557
40. Daniel BK (2019) What constitutes a good qualitative research study? Fundamental dimensions and indicators of rigour in qualitative research: the TACT framework. In: Proceedings of the European conference of research methods for business & management studies, pp 101–108

41. Vonk J (2019) Going digital: privacy and data security under GDPR for quantitative impact evaluation
42. Tuffour I (2017) A critical overview of interpretative phenomenological analysis: a contemporary qualitative research approach. J Healthc Commun 2:52
43. Chandra Y, Shang L (2017) An RQDA-based constructivist methodology for qualitative research. Qual Market Res: Int J
44. Shukla M, Johnson SD, Jones P (2019) Does the NIS implementation strategy effectively address cyber security risks in the UK?. In: 2019 international conference on cyber security and protection of digital services (Cyber Security). IEEE, pp 1–11
45. Basias N, Pollalis Y (2018) Quantitative and qualitative research in business & technology: justifying a suitable research methodology. Rev Integr Bus Econ Res 7:91–105
46. Porcedda MG (2018) Patching the patchwork: appraising the EU regulatory framework on cyber security breaches. Comput Law Secur Rev 34(5):1077–1098
47. Mohajan HK (2018) Qualitative research methodology in social sciences and related subjects. J Econ Dev Environ People 7:23–48
48. Yu HL (2020) Business as usual during an unprecedented time-the issues of data protection and cybersecurity. Contempor Asia Arbitr J 13(1):43–64
49. Mitchell EM (2020) Cyber security@ home: the effect of home user perceptions of personal security performance on household IoT security intentions (Doctoral dissertation, Syracuse University)
50. Liu N, Nikitas A, Parkinson S (2020) Exploring expert perceptions about the cyber security and privacy of connected and autonomous vehicles: a thematic analysis approach. Transport Res F: Traffic Psychol Behav 75:66–86
51. Gerke S, Shachar C, Chai PR, Cohen IG (2020) Regulatory, safety, and privacy concerns of home monitoring technologies during COVID-19. Nat Med 26(8):1176–1182
52. Riddel F (2020) Garmin grounded as ransomware cyber attacks pick up cadence. Govern Directions 72(8):380
53. Borkovich DJ, Skovira RJ (2020) Working from home: cybersecurity in the age of COVID-19. Issues Inf Syst 21(4)

An Investigation of Microarchitectural Cache-Based Side-Channel Attacks from a Digital Forensic Perspective: Methods of Exploits and Countermeasures

Reza Montasari, Bobby Tait, Hamid Jahankhani, and Fiona Carroll

Abstract In the current, fast paced development of computer hardware, hardware manufacturers often focus on an expedited time to market paradigm or on maximum throughput. This inevitably leads to a number of unintentional hardware vulnerabilities. These vulnerabilities can be exploited to launch devastating hardware attacks and as a result compromise the privacy of end-users. Microarchitectural attacks—the exploit of the microarchitectural behaviour of modern computer systems, is an example of such a hardware attack, and also the central focus of this paper. This type of attack can exploit microarchitectural performance of processor implementations, which in turn can potentially expose hidden hardware states. Microarchitectural attacks compromise the security of computational environments even within advanced protection mechanisms such as virtualisation and sandboxes. In light of these security threats against modern computing hardware, a detailed survey of recent attacks that exploit microarchitectural elements in modern, shared computing hard-

R. Montasari (✉)
Hillary Rodham Clinton School of Law, Swansea University, Singleton Park,
Swansea SA2 8PP, UK
e-mail: Reza.Montasari@Swansea.ac.uk
URL: https://www.swansea.ac.uk

B. Tait
School of Computing, University of South Africa, Unisa Science Campus,
Florida, Johannesburg 1709, South Africa
e-mail: taitbl@unisa.ac.za
URL: http://www.unisa.ac.za

H. Jahankhani
Information Security and Cyber Criminology, Northumbria University,
110 Middlesex Street, London E1 7HT, UK
e-mail: hamid.jahankhani@northumbria.ac.uk
URL: http://www.london.northumbria.ac.uk

F. Carroll
School of Technologies, Cardiff Metropolitan Univfersity,
Llandaff Campus, Western Avenue, Cardiff CF5 2YB, UK
e-mail: fcarroll@cardiffmet.ac.uk
URL: http://fionacarroll.eu/

ware were performed from a Digital Forensic perspective. It is demonstrated that the CPU (central processing unit) is an attractive resource to be targeted by attackers, and show that adversaries could potentially use microarchitectural cache-based side-channel attacks to extract and analytically examine sensitive data from their victims. This study only focuses on cache-based attacks as opposed to other variants of side-channel attacks, which have a broad application range. The paper makes three major contributions to the body of knowledge: Firstly in terms of the broadness of the scope of the analysis and a detailed examination of the means by which the data is analysed for performing side channel attacks, secondly with regards to how novel uses of data can facilitate side channel attacks, and thirdly also in the provision of an agenda for directing future research.

Keywords Microarchitectural analysis · Side channels · Digital forensics · Digital investigation · Attacks · Exploits · Countermeasures · CPU vulnerabilities · Cache-based attacks

1 Introduction

Hardware development is a cutting edge industry. Manufacturers must run a fine balance between the various requirements that are found in the modern hardware market. Hardware manufacturers are often, due to strong market competition, forced to focus on aspects such as development throughput, performance of a processor measured in performance per Watt and even being first to deliver a particular hardware feature to the market. A central, and very complex hardware component is the CPU. Although the deep rooted security of a processor is considered during the design phase of the CPU, certain hardware vulnerabilities can be exploited in the functioning of the processor in order to launch certain devastating hardware attacks on a processor, resulting in compromising the privacy of end-users using these computing systems. Microarchitectural attacks (MAs) is the consequence of such hardware vulnerabilities; it can exploit microarchitectural performance of various processor implementations, potentially exposing hidden hardware states such as cryptographic keys. For instance, an adversary can perform a differential power analysis [1–3] or monitor electromagnetic radiation [4] in order to deduce vital data from victims' systems [5]. MAs compromise the security of computational environments, even within advanced protection mechanisms such as virtualisation and sandboxes. To launch an MA, adversaries do not need to have elevated system privileges (all malicious activities are performed within their authorised privilege level), a knowledge of the plaintext or the ciphertext [5–7].

Furthermore, the performance of some processor components, create data-dependent variations (during the execution of cryptosystems) in terms of runtime and power consumption signatures. These variations can directly leak out the key value during a single cipher execution. The security vulnerabilities found in processor components are due to their design and implementation [8–11]. Although hardware manufacturers have been able to conceal the internal CPU architecture from programmers to a large extent, the CPU's timing behaviour is still highly visible. Data thefts exclusively by means of hardware have received substantial attention in the literature [2, 12]. Similarly, existing studies have scrutinised microarchitectural channels (MCs) from the perspective that MCs can be deployed to launch attacks against virtual machines (VMs) [13–18]. However, one research area that has remained relatively unexplored is data leakage at the interface between software and hardware [8, 19].

Therefore, in light of the discussion above, this study surveys cache-based side-channel attacks (CBSCAs) with a view to provide clarity with regard to appropriate countermeasures that will address the existing gap between current processor architectures and the required secure computing environment. It is demonstrated how cache architecture can be exploited and how instruction paths can be exposed by introducing various attacks that depend on the instruction cache (I-cache) architecture of CPUs. The study proceeds to propose new countermeasures followed by the analysis of the existing countermeasures in the related work. From the presented analysis, insight is deduced in relation to the current state of knowledge observed in the literature, hypothesise means of attack that have not been previously explored, predict possible future modus operandi of attacks, and propose effective future directions for the development of appropriate defence mechanisms. Although theoretical work of clear relevance is presented, the primarily focus of this paper is on practical, established attacks and relevant defence mechanisms. The paper makes three major contributions to the body of knowledge: Firstly in terms of the broadness of the scope of the analysis and a detailed examination of the means by which the data is analysed for performing side channel attacks, secondly with regards to how novel uses of data can facilitate side channel attacks, and thirdly also in the provision of an agenda for directing future research.

The remainder of the paper is structured in the following way: Sect. 2 presents CPU cache structure. Section 3 provides the findings of the survey, and a critical discussion of the work, presented in the paper. Section 4 proposes countermeasures to mitigate these attacks, paving the way for discussion of possible future work. Finally, the paper is concluded in Sect. 5.

2   CPU Cache Structure

A cache is divided into various fixed size blocks, called cache blocks or cache lines. Almost all modern CPUs have several levels of CPU caches with a split L1 cache, L2 caches and also, for larger processors, L3 caches. The L2 cache is rarely partitioned and functions as a common vault for the already partitioned L1 cache. Each core of a multi-core processor contains a separate L2 cache and is often not shared between the cores. In contrast, the L3 cache is shared amongst cores and is not partitioned as demonstrated in Fig. 1, adapted from [20]. An L4 cache is often on dynamic random-access memory (DRAM), as presented in Fig. 2, as opposed to static random-access

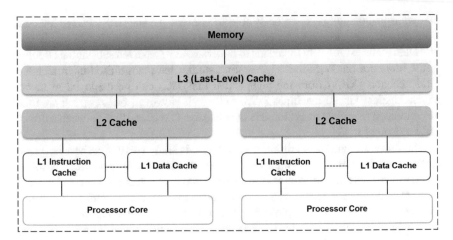

Fig. 1 Memory architecture in a modern x86 processor, adapted from [20]

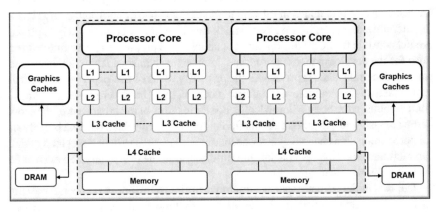

Fig. 2 DRAM representation of a processor accessed by a set of L4 tags included in the LLC (L3) of each core

memory (SRAM) on a separate die or chip. Each extra level of cache is often larger and can be augmented differently [21]. When the processor needs to read from or write to an area in the main memory, it first carries out an inspection to determine whether a copy of that data is in the cache. If the copy is residing in the cache, the processor instantly reads from or writes to the cache, which is considerably faster than reading from or writing to the main memory [22].

Almost all modern desktop and server CPUs contain at least three independent caches, including: an instruction cache to accelerate executable instruction fetch, a data cache to accelerate data fetch and store, and a translation lookaside buffer (TLB), used to accelerate virtual-to-physical address translation for both executable instructions and data. A single TLB could be provided for access to both instructions and data, or a separate instruction TLB (ITLB) and data TLB (DTLB). The data cache is often structured as a hierarchy of more cache levels (L1, L2, L3, etc.), as discussed above and shown in Figs. 1 and 2. Nevertheless, the TLB cache belongs to the memory management unit (MMU) and is not directly associated with the CPU caches [13]. The least volume of data that might be read from the main memory into a cache at a given time is considered to be the cache line or cache block size. This represents the movement of data between memory and cache in blocks of fixed size. Each cache miss causes a cache block to be recovered from a higher-level memory. The spatial locality property is the reason that a block of data is moved from the main memory to the cache as opposed to moving only the data that is currently required. Because a cache is restricted in size, placing new data in a cache dictates eviction of some of the data that was stored before [8, 23].

The process by which a cache line is copied from memory into the cache creates a cache entry which will consist of the copied data in addition to the required memory location [24, 25]. When a processor needs to read from or write to an area in the main memory, it first examines the main memory for a matching entry in the cache. The cache will then inspect the contents of the demanded memory area in any cache lines that are likely to have that address. If the processor determines that the memory area resides inside the cache, a cache hit has taken place [26, 27]. However, if no memory location is found inside the cache, a cache miss has occurred instead. If the cache hit has occurred, the processor will then instantly read or write the data from or into the cache line. However, if a cache miss has occurred, the cache will create a new entry and copy data from the main memory, and then the demand has been met from the contents of the cache [25, 28, 29].

The execution performance of a cache is affected by two elements: mapping structure and replacement policy.

2.1 Mapping Structure

The mapping structure defines the manner in which the lines are arranged within the cache. In contrast, the replacement policy, discussed in Sect. 2.1, determines which line should be evicted from the cache if an incoming call is to be placed in the cache. Cache mapping is a method by which it is determined where to store and search for data in a cache. There are three types of cache mapping techniques as represented in Fig. 3. These include: (1) direct mapped cache (DMC), (2) fully associative cache (FAC) and (3) N-Way set associative cache (NSAC). In a DMC, it is easy to check for a hit because a specific data block can be cached only in one single place. However, a DMC has a poorer performance since again there is only one location in which any address is cached. In a FAC, a data block is cached in any location which is dictated by the replacement policy. The FAC is capable of caching any address that needs to be stored [28, 30]. The reader should take note that the issue encountered by the DMC does not exist in the FAC since there is an allocated single line that

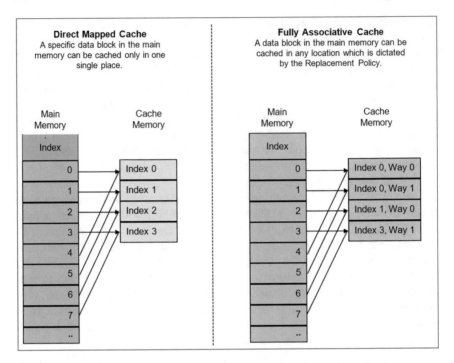

Fig. 3 Cache mapping structure of direct mapped cache and fully associative cache. Adapted from [33]

an address needs to use. The NSAC is an integration of the DMC and FAC. NSACs are partitioned into cache sets, each of which consists of the identical set number of cache lines [31, 32].

2.2 Replacement Policy

Although a data block can be placed only in some cache sets, it can be stored in any location within this block. Similarly, the specific place of data within its cache set is dictated by the replacement policy. The replacement policy is a technique to make room for the new entry on a cache miss, requiring the cache to evict one of the existing entries [34]. The fundamental issue with the replacement policy is that it will need to decide which cache entry is least likely to be utilised in the near future. There are various cache replacement policies (the discussion of which is beyond the scope of this paper; cite references to these instead so readers can find out more if beyond the scope of the paper) that have been suggested. However, the most commonly employed policy, least-recently used (LRU), replaces the entry that is least recently used between the entire data blocks that might be evicted from the cache. In relation to catch misses, they occur when data that is requested for processing by a component or application cannot be located in the cache memory. Thus, a failed attempt to read or write a piece of data into cache will result in execution delays, by requesting the program or application to fetch the data from other cache levels or in the main memory. For each new request, the processor looks for the main cache to locate that data. If the data is not located, it will then be seen as a cache miss. Each cache miss causes the overall process to slow down since the CPU will look for a higher-level cache including L1, L2, L3 and random-access memory (RAM) for that data. Moreover, a new entry is made and placed in a cache prior to the processor being able to access it. There are three types of cache misses including: (1) instruction read miss, (2) data read miss, and (3) data write miss. Cache read misses from an instruction cache often create the longest delay since the processor is forced to wait until the instruction is fetched from the main memory. Cache read misses from a data cache often create a shorter delay as instructions not dependent on the cache read can be generated and continue execution until the data is returned from the main memory, and the dependent instructions can resume execution. Cache write misses to a data cache often create the shortest delay since the write can be kept in queue, and there are few restrictions on the execution of ensuing instructions. The processor can carry on until the queue is full. Many processors employ different caches for data and code segments of a process.

In light of the discussion above, different types of CBSC exploits are described in the next section prior to suggesting various countermeasures in Sect. 4.

3 Findings of the Survey and Discussion of the Findings

3.1 Cache-Based Side-Channel Attacks

A CBSCA enables an unprivileged process to attack another process running side by side on the same processor. This enables the adversary to extract sensitive data from the victim through shared CPU caches. This sensitive data is often related to both cryptographic processes such as signing or decryption [15, 35]. It is emitted via secret-dependent data flows that result in cache usage patterns that are visible to an adversary. With knowledge of this, an attacker can leverage various methods to manipulate data in the shared cache to infer the victim's cache usage patterns. As a result, the adversary will be able to deduce sensitive data that prescribes such patterns. A CBSCA can also be carried out by monitoring operation such as AES T-table entry or modular exponentiation multiplicand accesses [101]. The memory accesses of software cryptosystems use key-dependent table lookups. Exposing such memory access patterns through cache statistics and the knowledge of the processed message renders it possible to exploit these ciphers. Moreover, there are certain CBSCAs that can be conducted remotely over networks. Within a cloud computing environment, CBSCAs can facilitate cryptographic/bulk key recovery. A cloud's shared resources such as CPU and memory can provide the attackers with side-channels which leak sensitive data and facilitate key recovery attacks. Those users that operate outdated libraries that are susceptible to leakage are more vulnerable to mass surveillance [36, 37].

The most commonly exploited leakage in the shared resource systems derive from the cache and the memory. An RSA secret key from a co-located instance can be obtained by conducting a Prime+Probe attack [16, 38, 39]. In this situation, to accelerate the attack, an adversary with advanced programming skills will be able to reverse engineer the cache slice selection algorithm, for instance for Xeon (a brand of x86 microprocessors targeted at the non-consumer workstation, server, and embedded systems) that is employed in distributed systems such as cloud computing. Furthermore, the attacker can also use noise reduction to infer the RSA private key from the monitored traces. By processing the noisy data, they will be able to retrieve the RSA key used during decryption as shown in [9, 40]. Based on the existing studies, we categorise cryptanalytic CBSCA into three classifications: trace-driven, time driven and a new category, access-driven, in accordance with the type of information that the adversary discovers about a victim's cipher. As previously discussed, in trace-driven attacks, the attacker monitors the sequence of cache hits and misses during a cryptographic cipher execution. By observing the execution details of the cipher, the adversary will be able to extract important information from specific key-dependent memory access results within a cache hit or a cache miss. This denotes that trace-driven attacks enable the adversaries to discover the result of each of the victim's memory accesses in relation to cache hits and misses [41, 42]. They are often carried out in and against hardware because of the difficulty of obtaining the trace of cache hits and misses in software.

In time-driven attacks, adversarial parties will monitor the entire runtime of a cryptographic cipher execution. They will then be able to extract important information since the runtime relies on the number of key-dependent memory accesses which lead to cache misses. Because the cache performance is only one part of the many aspects that impacts the total runtime of a cryptosystem, time-driven attacks would require statistical analysis using a large number of samples to deduce important information. Access-driven attacks monitor partial information of the addresses that the victim accesses. Similar to time-driven attacks, the timings of the cache are examined as a source of information leakage. Access-driven attacks probe the cache behaviour with greater detail as opposed to assessing the overall runtime [43, 44]. They are capable of identifying whether or not a cache line has been ejected as the main technique to launch an attack. Access-driven cache attacks themselves, can be classified into three subcategories, including: Evict+Time, Prime+Probe, Flush+Reload. Whilst many CBSCAs use one of these three methods, there exist other variations to match the specific abilities of hardware and software environments [7, 45]. The remainder of this section analyses cache manipulation techniques used to carry out CBSCAs.

3.2 Attacks

Evict+Time

In this attack, the adversary fills a cache set with his own lines, then waits for a specific period and proceeds to establish whether the lines are still cached.

Prime+Probe

In this attack, the adversary fills a cache set with his own lines, then waits for a specific period and proceeds to establish whether the lines are still cached. The adversary will then be able to determine whether the victim accessed the designated cache set in the meantime. This requires that the attacker examines certain cache sets to establish the presence of a cache miss. To do so, he will need to time the accesses to the cache set after the victim executes [46], [?,?,?]. They assign a group of cacheline-sized, cacheline-aligned memory blocks in order for such memory blocks to fill a collection of targeted cache sets. Having done this, the attacker will then constantly carry out two attack phases including 'prime phase' and 'probe phase'. Within the 'prime phase', the attacker examines each memory chunk to eject all the victim's data in such cache arrays. They will then wait for a delay time prior to carrying out the 'probe phase', where each memory chunk is examined in the group again and the time of memory accesses is determined [45, 47]. Longer access times represent one or more cache misses. This means that this cache array has been accessed by the victim between the prime and probe phases. The attacker will repeat these two phases many times in order to acquire traces that might overlap with the victim's performance of cryptographic operations [13, 44]. As a result, the adversary will be able to establish which cache lines were replaced by the victim and infer more details concerning which addresses the victim accessed. To determine the cache lines that were replaced, the attacker will need to measure the speed of each cache access. It should be noted that the Prime + Probe attacks can target both the L1 cache [38, 48], as well as the LLC [14, 49].

Adversaries can also attack AES (See Sect. 6 for details) in Open SSL 0.9.8 with Prime+Probe on the L1 data-cache (D-cache) [7, 50] and L1 instruction cache (I-cache) contention to establish an end-user's control flow. This will enable the attacker to differentiate squares and multiples in OpenSSL 0.9.8d RSA [19, 51]. Prime+Probe attack can also be carried out within cloud computing environments. The cross-VM emission exists in public clouds and is often a practical attack vector for stealing sensitive data [37, 52]. In the cross-VM context, the adversary and victim have two distinct VMs running as co-tenants on the same server. Thus, the adversary will be able to acquire co-tenancy of a malign VM on the same server as a target [14, 53]. For instance, using a Prime+Probe technique, a cross-VM attack can be carried out to obtain ElGamal secret keys from the victim [7, 54].

Cache Template Attacks

CBSCAs attacks can also be carried out through cache template attacks (CTA) based on the application of the Flush+Reload attack. A CTA is a generic technique that can be performed online on a remote system without any prior offline computation, and comprises of two stages including a 'profiling phase' and 'exploitation phase' [55, 56]. Within the profiling phase, the attacker establishes dependencies among the processing of secret information such as private keys of cryptographic primitives and specific cache accesses [56–58]. Within the exploitation phase, the attacker extracts the secret values based on observed cache accesses. A CTA can enable adversaries to profile and take advantage of cache-based information emission of any program automatically without the advance knowledge of specific software versions [55]. One of the detrimental consequences of such an attack is the acquisition of keystrokes by the adversaries or the identification of specific keys on an operating system's user interface. Since the LLC is physically indexed and tagged, building a prime buffer for a Prime+Probe attack necessitates knowledge of virtual to physical address mappings [13, 19]. Evict+Reload, a technique that integrates the eviction process of Prime+Probe with the reload step of Flush+Reload, is slower and less accurate than Flush+Reload [45, 57]. However, adopting such a technique removes the need for dedicated instructions for flushing cache lines.

Flush+Reload

The Flush+Reload attack, itself, is a variant of Prime+Probe attack, that is based on the assumption that identical memory pages can be shared amongst malicious and victim processes. To carry out a Flush+Reload attack, the adversary and victim will need to share memory pages through which a particular memory is ejected from the entire cache hierarchy. Hence, a Flush+Reload attack exploits physical memory that is shared between the attacker and the victim security domains, in addition to the capability to eject such pages from LLCs, employing an ability similar to that offered by the Clflush instruction on the x86 architecture [39, 59]. Operating systems use shared memory pages among processes running on the system to reduce the memory footprint of a system. Libraries employed by multiple programs are mounted on physical memory only once and are then shared between the processors employing it [47, 57, 60]. Otherwise the sharing can be implemented by searching and merging identical contents. When splitting a process, the memory is initially shared between the parent and child processes. After processes write into a shared memory slice, a copy-on-access page fault is created, and the operating system generates a copy of the matching memory region. This denotes that in order to retain the isolation among non-trusting processes, the system needs to depend on hardware structures that impose read only or copy-on-write 'semantics' for shared pages. The shared application of the processor cache can lead to one method of interference via shared pages. Once a shared page in memory has been accessed by a processor, the content of the location of the memory that has been accessed will be cached.

To carry out the Flush+Reload attack, the adversary selectively flushes a shared line from the cache, and after waiting for a certain period of time, will then examine whether it was fetched via the victim's execution. This method could potentially be used to reduce the memory footprint of multi-tenanted systems [15, 61]. Therefore, the malicious process can employ this to observe access to a memory line. Because of certain flaws in processors such as Intel X86, page sharing can subject processes to information leaks [15, 58]. Hence, a Flush+Reload attack can leverage such flaws to monitor access to memory lines in shared pages. For instance, a Flush+Reload attack can be launched against the L3 on processors with Last Level Cache (LL3) by employing either memory mapping or page de-duplication. This will result in the attack program and the victim not needing to share the execution core. It can be employed to steal private encryption keys from a victim program [38, 48, 57]. Often, the Flush+Reload is carried out in two phases including 'flush phase' and 'reload phase'. In the flush phase, the adversary extracts the designated blocks out of the cache hierarchy through the Clflush instruction, before waiting for a set inter-mission where the victim releases the vital directions, after which they are fetched back to the caches. In the reload phase, memory blocks are reloaded into the caches to determine the access time. A short access time for one memory block represents a cache hit. This denotes that this block has been accessed by the victim during the intermission. The attackers can then extract traces of the victim's memory accesses and infer the sensitive data by repeating these two phases [44, 45, 57]. Flush+Reload attacks have been extensively examined in the literature. Various studies have been carried out in relation to different variations of Flush+Reload attacks [44, 45, 57]. This consists of various elements of the CPU's branch-prediction facility [8, 45, 57, 62], the DRAM row buffer [56, 63], the page-translation caches [64–66] and also other microarchitectural aspects [45, 62].

Memory Deduplication

Memory deduplication is a method to lessen the memory footprint across virtual machines. It has now also become a default feature of Microsoft Windows 10 Oper-ating Systems. Deduplication maps several matching data blocks of a physical page onto a single shared data block with copy-on-write semantics. Consequently, a write to such a shared page activates a page fault and is therefore slower than a write to a standard page. An adversary who will be capable of creating pages on the target system will be able to employ this timing difference as a single-bit side channel to learn that certain pages reside in the system. Deduplication side channels can provide an adversary with a machine to read arbitrary data in the system. An attacker who controls the alignment and reuse of data in memory will be capable of carrying out byte-by-byte exposure of sensitive data including randomised 64-bit pointers). He will still be able to expose 'high-entropy' randomised pointers even without control over data alignment or reuse. An end-to-end JavaScript-based attack, which is based on a combination of Deduplication and the Rowhammer hardware vulnerability, can also be launched against a web browser to extract arbitrary memory read and write access in the browser [10, 36, 38].

4 Solutions and Recommended Countermeasures

4.1 Advanced Encryption Standard Algorithms

Advanced encryption standard (AES) algorithms in hardware could be used as a countermeasure to safeguard modern hardware against CBSCAs. Few manufacturers have started implementing better hardware support in the design of their processor technologies to offer better constant-time cryptography operations. For instance, Intel has introduced AES new instructions (AES NI), which is a new encryption instruction set that enhances the AES algorithm and speeds up the encryption of data in the two categories of Intel Xeon processor and the Intel Core processor. Thus, AES-NI could potentially provide an advantage in relation to speed over other implementations. Moreover, since AES-NI, which consists of seven new instructions, was specifically developed to be constant-time, it provides a better protection against CBSCAs over some other software implementations. The purpose of AES instructions is to alleviate all known timing and CSC emission of sensitive data from Ring 3 spy processes. Their latent period is data-independent, and because all the computations are carried out internally by the hardware, there will be no need for lookup tables.

As a result, if AES instructions are employed properly, the AES encryption/decryption and the key expansion will then have data-independent timing and involve only data-independent memory access. Therefore, the AES instructions will facilitate writing high performance AES software which is simultaneously safeguarded against the currently known software CBSCAs. However, in relation to the timing side channel attacks (the discussion of which is beyond the scope of this paper), C-like functions can also be used as a defence mechanism to take advantage of AES-NI opcodes which are constant-time (i.e. opcode and throughput and latency are fixed). In this context, this is a more effective countermeasure over the software implementations, on condition that (1) the opcodes are properly employed (e.g. performing the cipher blocker chaining decryption in tandem whilst handling the AES itself in a constant-time manner), and (2) the previously mentioned software implementations are not constant-time. AES-NI opcodes are effective in that they can offer reasonable performance (even though its codes in CBC decryption and CRT implementations can be accelerated more) and better protection against CBSCAs.

Despite its many benefits, AES-NI, itself, can, at the same time, facilitate a CBSCA in certain situations. For instance, it is relatively easy to build backdoors into a CPU that will execute arbitrary code when a specific data pattern is activated. This will be difficult to detect in the hardware and can be remotely activated if the attacker manages to inject data into the compromised CPU, for instance by transmitting an IP packet. However, this backdoor is only an entrance and needs a payload to exploit actively. This denotes that the real exploit is executed in software and can easily be discovered. It will leave traces of an exploit and needs the payload to operate on an unknown machine. However, the AES-NI instructions facilitate a new targeted attack. For instance, by adapting a chip to store the last 'n' keys placed in AES-NI instructions, it would be possible for an attacker to have an almost undetectable key escrow (an arrangement in which the keys required to decrypt/encrypt data are held in escrow so that, under certain situations, an authorised party might gain access to those keys), enabling him to access the key later (for instance, by employing a specific data pattern that is replaced with the key). This will be a quiet attack and have the potential to work until it is finally detected. Notwithstanding this shortcoming, we believe that AES-NI is currently one of the best software implementations for AES to provide protection against CBSCAs (even though soon there might be faster software implementations for AES that can be utilised by more effective cryptography libraries). However, the degree to which AES-NI can provide protection against CBSCAs is not obvious yet due to the lack of research in this area.

Following the recommended countermeasures above against CBSCAs, it could also be possible to increase the runtime by injecting dummy operations within AES algorithms. With adequate dummy loads injected, it will be very difficult to determine whether a specific cache-hit or cache-miss is generated by genuine or false execution. Although injecting dummy operations will modify the runtime randomly, at the same time, it will be independent of cache activities. This approach must be considered only in cases where attacks are carried out based on behaviour traces as opposed to the timing information.

4.2 Address Space Layout Randomization

Address space layout randomization (ASLR) is another defence method that can be used to safeguard both user space and kernel space code. ASLR is capable of placing address space targets in random locations that cannot be guessed by the attackers. If an adversary tries to take advantage of an incorrect address space location, the target application will crash, preventing the attack and warning the system. Figure 4 demonstrates the manner in which ASLR functions by randomising various locations of Windows' critical components in memory between restarts. ASLR was developed by the Pax Project as a Linux patch in 2001 and was subsequently built into the Windows OS in 2007. Before the implementation of the ASLR, the memory locations of files and applications could easily be determined. The correct execution of the ASLR can significantly diminish certain attacks and associated exploitation techniques. ASLR randomises the system's virtual memory layout when new code execution initiates or when the system is loaded. Consequently, this randomisation prevents the attacker from identifying the virtual address of applicable memory locations required to carry out a control-flow hijacking attack, where attackers intend to take advantage of the susceptibility of a memory corruption. By utilising the ASLR, it would be possible to prevent sophisticated exploitation techniques such as return-into-libc attack and return oriented programming (ROP). ASLR has been incorporated into the operating systems of many modern computing devices such as desktops, servers and mobile phones [60, 65, 67].

However, kernel space ASLR still has limitations against local attackers with restricted privileges [64]. For instance, an attacker might be capable of executing a

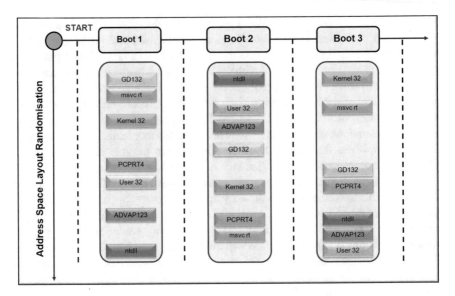

Fig. 4 ASLR functions by randomising various locations of Windows' critical components in memory between restarts. Adapted from [34]

generic CBSCA against the memory management system to infer information about the privileged address space layout. Another example is that an attacker could implement BTB crashes among their branch instructions process either on the user-level victim process or on the kernel running on its behalf [62]. Such crashes can affect the timing of the adversary's code, enabling places of known branch instructions to be detected in the address space of the victim's process or kernel. This can potentially lead to the attacker retrieving the kernel ASLR values. Brute-force and memory disclosure attacks can also be carried out against ASLR. Furthermore, certain properties of Intel CPUs can also be exploited to identify which memory accesses are allocated [65, 67].

4.3 Signature-Based and Anomaly-Based Techniques

One method to detect and mitigate CBSCAs is to implement a method that integrates both signature-based detection techniques (SBDTs) and anomaly-based detection techniques (ABDTs). If implemented properly, SBDTs would be able to identify when a safeguarded system executes a cryptographic application, while the ABDTs could monitor protected systems to detect abnormal cache behaviours during CBSCAs. SBDTs could potentially address both Prime+Probe and Flush+Reload CBSCAs, which signify the entire identified LLC side channels in modern computer systems [39, 44, 48]. Similarly, SBDTs could be utilised to identify network intrusion and malware by comparing monitored application or network features with pre-detected attack signatures. Likewise, to identify CBSCAs, signatures of side-channel attacks need to be created from the entire known CBSCA techniques and employed to compare with events acquired from production systems. In this regard, classification algorithms can be used to distinguish standard programs from Prime+Probe attack programs.

Although this approach has significant benefits in that it provides a high true positive rate in identifying known attacks, at the same time, it has some drawbacks. One of the drawbacks of this method is that it could potentially be bypassed by savvy attackers by altering the memory access pattern in a Prime+Probe attack to avoid signature-based detection. Analogous to SBDTs, the ABDTs could be used to detect network intrusion and misuse by monitoring system activity and classifying it as either normal or anomalous [68]. Such classification is centred on heuristics as opposed to patterns or signatures and aims to identify any type of misuse that does not correspond to normal system operation. This is in contrast with SBDTs that can identify attacks for which a signature has been generated beforehand [69, 70]. Within network systems, two methods can be used to detect anomalies including: (1) employing artificial intelligence techniques [71], and (2) establishing what constitutes a normal usage of the system by utilising a strict mathematical model and identifying any divergence from this as an attack. There exist other techniques for detecting anomalies such as data mining-based methods, grammar-based methods, and artificial immune system-based methods, the discussion of which is beyond the scope of this work.

In the context of computing hardware or computer end points (relevant to this study), host-based anomalous intrusion detection systems (HBAIDSs) are implemented as one of the last layers of defence [72]. These HBAIDS enable granular protection of end points at the application level [71]. In anomaly-based detection, the typical activities of harmless applications could be modelled and any divergence from such models identified as potential attacks [71, 73]. To identify CBSCAs using such methods, models for harmless application activities could be developed. The application activities could then be monitored to determine whether they comply with the models in the database. In contrast with SBDTs, ABDTs can detect 'zero-day' attacks as well as the known attacks [68, 74]. However, similar to SBDTs, there are also drawbacks associated with ABDTs. For instance, such techniques could have a high false-positive rate and the ability to be misled by a correctly delivered attack. Another drawback of employing the ABDTs to address CBSCAs derives from the difficulty of accurately modelling harmless application behaviour by means of performance counters [68, 74].

4.4 Hardware Performance Counters

Hardware Performance Counters are sets of registers often built into modern computing hardware that monitor various performance activities such as the number of cache evictions or branch mispredictions taking place on a certain core. They are contained in x86 (e.g., Intel and AMD) and ARM processors in order to store the counts of activities associated with hardware within computer systems [138, 229]. An adversary (local attacker) who has access to such perform counters might be able to deduce the sensitive details utilised during a protected function's performance [44, 45, 75–77]. Therefore, a Hardware Performance Monitoring Counter, presented in Figs. 5 and 6, can be utilised to restrict access to performance monitoring counters in order that a user's process is not able see detailed performance metrics of another user's processes. These counters operate in conjunction with event selectors to determine certain hardware events that carry out low-level performance analysis and update a counter after a hardware event takes place [45, 69]. For instance, the Intel® Performance Counter Monitor is used as a method to measure the CPU utilisation [76] as represented in Figs. 5 and 6. Available hardware counters in a processor are often restricted, and each given CPU model is likely to have various events that need to be measured. A specific counter can be programmed with the index of an event type to be observed such as a L1 cache miss. A large number of modern processors offer a Performance Monitor Unit (PMU), allowing applications to monitor performance

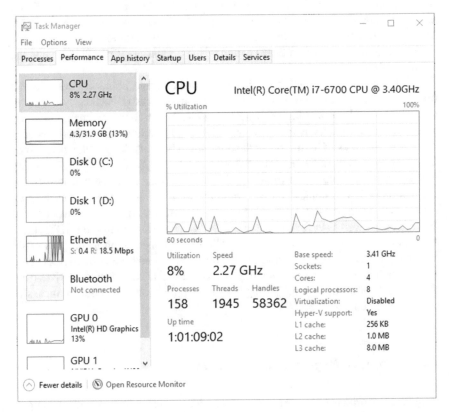

Fig. 5 Window's 10 Intel® performance counter monitor used to measure the CPU utilisation

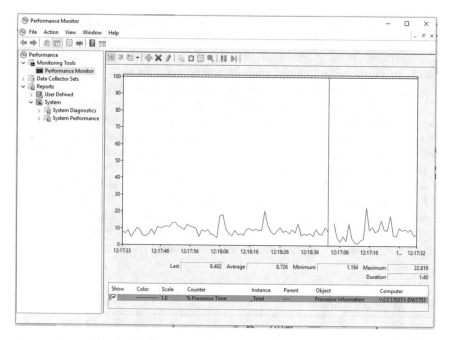

Fig. 6 Windows 10's Intel® performance monitor showing data from performance counter monitor

counters, as discussed above [44, 45]. An instance of PMUs is the Interrupt-Based Mode, under which an interrupt is caused when the incidences of a certain event surpass a predefined threshold or when a predefined amount of time has passed. This facilitates event-based sampling and also time-based sampling. Performance counters are capable of divulging programs' execution features that can further disclose the programs' security conditions.

4.5   Isolation

The sharing of the caches between adversary and victim can be exploited as side-channels in external interference attacks. One method to prevent information leakage is to stop cache sharing between the adversary and victim by separating the cache into various zones for different purposes [13, 15, 78]. The uthor in [5] proposed two cache designs including a Static Partitioning (SP) cache and Partition-Locked (PL) cache. The SP cache is statically split into two sectors either through ways or sets. In SP caches, divided through ways, each given way is kept for either the victim or the attacker program. In the same way, in SP caches partitioned through sets, each specific set is also put aside for either the victim or the attacker program.

Because of the removal of cache line sharing, SP caches are capable of stopping external interference. However, this has a negative impact on performance due to the static cache partitions. PL caches employ a more detailed dynamic cache partitioning policy. In PL cache, each given memory address contains a defence bit to signify whether this memory block requires locking in the cache. After the safeguarded block has been locked in the cache, it cannot be substituted by a defenceless block (the attacker's data). Rather, the unprotected block will be directed among the processor and the memory without making the cache full. This substitution counteracts the adversary's scheme to spy on the victim's cache accesses. If confidential data is pre-loaded prior to encryption initiating, then a PL cache allows constant-time memory accesses because all accesses to confidential data will lead to cache hits [5]. Although isolation techniques might help to prevent certain CBSCAs within conventional settings, their protection against side-channels within resource sharing systems is limited. Even with sophisticated isolation methods, resource sharing still presents significant security threats to cloud customers [37, 53].

4.6 Randomisation

Similar to ASLR (see 4.2), in this method, randomisation could be incorporated into the adversary's measurements in order to render it difficult for the attacker to acquire accurate information based on monitoring [44]. This consists of random memory-to-cache, mappings, cache prefetches [79], timers [49], and cache states [80]. The randomisation method would involve randomising side-channel information. As a result, no correct information would be leaked out of caches [47, 81]. Randomisation methods could be classified into three main categories, including: (1) Random-Eviction (RE) cache, (2) Random-Permutation (RP) cache [42, 82], and (3) NewCache [42]. The RE cache method would entail the inclusion of random noise in the adversary's observation and intermittently selecting a random cache line to eject [5]. This could potentially prevent the adversary from being able to determine whether an observed cache miss is the result of the cache line substitution or the system's random ejection. This would add to the adversary's technical challenge to attain hidden information such as a cipher key. The RP cache method could enable randomising the mappings from memory addresses to cache by utilising a 'Permutation Table' (PT) for each specific process. This would allow a dynamic mapping from memory addresses to hardware-remapped cache sets. Because the victim system's memory-to-cache mappings are dynamic and unidentified to the adversary, he would not be able to establish which lines the victim actually accessed. This is in contrast with traditional caches which contain both static and fixed memory-to-cache mappings as opposed to dynamic and randomised mappings. Similar to the RP cache approach, the NewCache method could be employed to randomise the memory-to-cache mappings. However, this Randomisation is based on a Logical Direct-Mapped Cache (LDM), that does not physically exist [5, 83] and can improve protection against information leakage.

4.7 Restriction of Clflush Instruction

There is an amalgamation of four elements that render the Flush+Reload attacks possible, including: (1) data flow from sensitive data to memory access patterns, (2) memory sharing among the adversary and the victim, (3) precise, high-resolution time measurements, and (4) the unrestricted use of the Clflush instruction. Therefore, the Flush+Reload attacks could be prevented by obstructing any one of these four elements. The Clflush command available on user space (userland) for x86 devices can be utilised as a countermeasure to eject cache lines associated with the aggressor row addresses among its memory accesses [11]. The X86 architecture does not provide permission checks for using the Clflush instruction; as a result, a key mitigation to the issue could be to restrict the control of the Clflush instruction. The important application of a Clflush instruction would be to impose memory coherence. This is a required condition where conforming memory locations for each processing element could include the same cached data [15, 61]. A Clflush instruction is required when using systems that do not support memory coherence. Moreover, Clflush is also employed to monitor the application of the cache for optimising a program. It is also used to limit the application of Clflush to memory pages where the process has write access and also to memory pages to which the system permits Clflush access. Such access controls could be executed by installing memory types that would limit flush access to the Page Attribute Table as described in Intel's Developer's Manual. Similar to any other types of CBSCAs, one of the factors that facilitates a Flush+Reload attack is the availability of a high-resolution clock. Consequently, lessening the clock resolution or presenting noise to a clock measurement could be used as a solution against the attack [15]. However, one should note that this method is limited in that adversaries could potentially employ other techniques to create high resolution clocks. For instance, they could utilise data from the network or execute a clock procedure in a different running core.

4.8 Other Recommended Solutions

Reordering memory access could also be an effective countermeasure whereby the random reordering of memory access would decrease the connection among a captured behaviour trace (or runtime) and the input and algorithm. To accomplish this, one would need to employ a non-deterministic processor architecture. Other methods that we suggest to avert the loss of information via a side channel breach of the cache performance include turning off cache S-Box access as opposed to the traditional method of removing the entire cache. It should, however, be noted that eliminating the entire cache could remove all instructions that rely upon a cache. Thus, it is important to remove only instructions that access cache memory. Turning off cache S-Box access, which uses cache-bypass in order to boot data directly from memory, could also be an effective option even though it might reduce performance due to cache-hits and cache-misses. We also propose the avoidance of lookup tables and employing some types of computed non-linear transformation in their place. This would facilitate the possibility of parallel implementation of this transformation denied by the demand for successive memory access. One could also use as a countermeasure hardware partitioning that is based on deactivating hardware threading, page sharing, presenting hardware cache partitions, quasi-partitioning, and migrating VMs within cloud services.

Reduction of resource sharing has been suggested as a countermeasure against CBSCAs within multi-tenant systems. This approach can be achieved by hardware alterations or by dynamically segregating resources [45, 84]. Cache cleansing techniques [80] can be employed to eliminate data leakage from time-shared caches. Similarly, cache colouring techniques that rely on time-based isolation via scheduling units and resource segregation could also be used to mitigate the occurrence of parallel side channels [52] or to separate various tenants within cloud environments [85]. It might also be possible to reduce access-driven and Flush+Reload CBSCAs that exploit last-level caches (LLCs) shared across cores to leak information. This approach could potentially address physical memory pages shared among security domains to deactivate the sharing of LLC lines, hence stopping access-driven and Flush+Reload CBSCAs through LLCs [61]. Another potential solution would be to employ Intel CAT to divide the LLC, circumventing the essential resource sharing that is exploited during many attacks [51]. Yet another mitigation method could be to identify CBSC emissions through the examination of 'static source code' or by carrying out dynamic anomaly identification using CPU performance counters [45, 57]. Furthermore, microarchitectural hardware-based techniques could be deployed to enable information flow tracking by design, in which non-interference would be employed as a baseline confidentiality property [86].

Moreover, non-monopolizable caches could be utilised [87] as a hardware protection against access-based attacks by placing a limit on the number of lines in each cache set that could be used by a process. Based on the extent of non-monopolization, an attacker would not be able to eject any of the victim's information from the cache, which removes access-based attacks. Partitioning caches could be another potential countermeasure which can be used within cloud computing environments to prevent cache sharing. This would entail separating cache memory into different zones by Surface-Enhanced Raman Spectroscopy (SERS) methods via various VMs [51, 83, 87], and software could achieve such a task. Additionally, in order to lessen the co-location probability among victim's and attacker's VMs in cloud computing environments, new VM placement policies have also been developed. This denotes that the VMs are frequently migrated to increase the difficulty of VM co-location for adversaries. Some studies have suggested employing an operation time delay feature as a method to counteract CBSCAs [11, 19]. This method would rely upon rendering the security operation time delay constant (or random), irrespective of the microarchitecture components used. However, we argue that this method would be ineffective as implementing constant-time execution code can be very difficult since optimizations presented by the compiler must be circumvented. One way to address this and make such an approach practical would be to develop and implement dedicated constant-time libraries in order to enable safeguarding systems against CBSCAs.

Based on the extent of non-monopolization, an attacker will not be able to eject any of the victim's information from the cache, which removes access-based attacks. Partitioning caches is another countermeasure which can be used within cloud computing environments to prevent cache sharing, by separating cache memory into different zones by SERS (Surface-Enhanced Raman Spectroscopy) methods via various VMs [51, 83, 87], and software can achieve such a task. Furthermore, in order to lessen the co-location probability among victim's and attacker's VMs in cloud computing environments, new VM placement policies have also been developed. This denotes that the VMs are frequently migrated to increase the difficulty of VM co-location for adversaries. Some studies have suggested employing an operation time delay feature as a method to counteract SCBAs [11, 19]. This method relies upon rendering the security operation time delay constant (or random), irrespective of the microarchitecture components that are employed. However, we argue that this method is ineffective as implementing constant-time execution code can be very difficult, since optimizations presented by the compiler must be circumvented. One way to address this and make such an approach practical is to develop dedicated constant-time libraries, that will need to be implemented in order to enable safeguarding systems against CBSCAs.

Bitslice implementation of the AES [88] could be another countermeasure which does not employ any lookup tables, preventing information from leaking out via cache side channels. Another mitigation method would be to carry out cache warming or pre-fetching [42]. Time-Driven and Trace-Driven CBSCAs distinguish between cache-hits and misses. Thus, eliminating this distinction has been suggested as a robust countermeasure [41]. To stop information from leaking, one would need to warm up the cache into which the lookup tables must be loaded prior to the runtime being initiated. In this situation, no cache misses would occur on condition that data was loaded to cache prior to runtime. As a result, there would be no leakage of data. Alterations to hardware-based circuits might also be able to prevent CBSCAs, the purpose of these countermeasures being to render the emission trace of O1 identical to the leakage trace of O0 [11].

5 Conclusion and Future Work

The existing studies on CBSCAs demonstrate various methods of extracting information from computing devices. CBSCAs have been shown to be the most difficult form of MAs to address without affecting the system performance. The extensive study presented in this paper highlighted the essential characteristics of the processors' execution units that cause side-channels, introduced and analysed various CBSCAs, and examined and proposed various countermeasures. From this work, it can be deduced that although there exist a few studies proposing advanced countermeasures against the attack vectors discussed in this article, these studies provide an inadequate understanding of the effectiveness of these countermeasures. Therefore, independent assessments of these countermeasures will need to be carried out to determine the efficacy of these defence mechanisms against a given attack vector, and to determine whether a mitigation mechanism can be bypassed or not.

Considering the discussion above, as a direction for future research, various exploitation techniques should be developed to establish the reliability of the proposed countermeasures present in the state-of-the-art, and then test them in a 'targeted' and 'predictable' manner within a dedicated forensic laboratory. For instance, such a study could examine Rowhammer exploitation methods that are highly likely able to compromise the existing isolation mechanisms by flipping bits in userspace binaries. This technique could take advantage of the system-level optimizations and a side channel to force the OS to place target pages at a location that we, as hypothetical attackers, would select. Through this technique, one could essentially attempt to exploit the CPU code instructions (which enables user-level code to assign private regions of memory that are safeguarded from processes running at higher privilege levels) so as to conceal his attack from both the end-user and the OS, rendering the detection of the attack impossible. We propose that it is only by conducting this technical study that one can truly provide adequate insight on the usefulness of the existing countermeasures and subsequently propose a way forward. In the interim, this study is a point of reference for both research communities and industry to advance security in the emerging field of Microarchitectural Analysis.

References

1. Kocher PC, Rohatgi P, Jaffe JM (2017) Secure boot with resistance to differential power analysis and other external monitoring attacks. US Patent 9,569,623
2. Kocher P, Jaffe J, Jun B, Rohatgi P (2011) Introduction to differential power analysis. J Cryptogr Eng 1(1):5–27
3. Schramm K, Leander G, Felke P, Paar C (2004) A collision-attack on aes. In: International workshop on cryptographic hardware and embedded systems. Springer, pp 163–175
4. Homma N, Aoki T, Satoh A (2021) Electromagnetic information leakage for side-channel analysis of cryptographic modules. In: IEEE international symposium on electromagnetic compatibility. IEEE, pp 97–102
5. Zhang T, Lee RB (2014) Secure cache modeling for measuring side-channel leakage. Technical Report. Princeton University
6. Gullasch D, Bangerter E, Krenn S (2011) Cache games—bringing access-based cache attacks on aes to practice. In: 2011 IEEE symposium on security and privacy. IEEE, pp 490–505
7. Osvik DA, Shamir A, Tromer E (2006) Cache attacks and countermeasures: the case of aes. In: Cryptographers' track at the RSA conference. Springer, pp 1–20
8. Aciiçmez O (2007) Yet another microarchitectural attack: exploiting i-cache. In: Proceedings of the 2007 ACM workshop on Computer security architecture. ACM, pp 11–18
9. Acıçmez O et al (2009) Microarchitectural attacks and countermeasures. In: Cryptographic engineering. Springer, pp 475–504
10. Bosman E, Razavi K, Bos H, Giuffrida C (2016) Dedupest machina: memory deduplication as an advanced exploitation vector. In: 2016 IEEE symposium on security and privacy (SP). IEEE, pp 987–1004

11. Fournaris A, Fraile LP, Koufopavlou O (2017) Exploiting hardware vulnerabilities to attack embedded system devices: a survey of potent microarchitectural attacks. Electronics 6(3):52
12. Wiley J (2008) Security engineering: a guide to building dependable distributed systems, 2ed Editio, pp 239–274
13. Liu F, Yarom Y, Ge Q, Heiser G, Lee RB (2015) Last-level cache side-channel attacks are practical. In: 2015 IEEE symposium on security and privacy. IEEE, pp 605–622
14. Ristenpart T, Tromer E, Shacham H, Savage S (2009) Hey, you, get off of my cloud: exploring information leakage in third-party compute clouds. In: Proceedings of the 16th ACM conference on computer and communications security. ACM, pp 199–212
15. Yarom Y, Falkner K (2014) Flush+ reload: a high resolution, low noise, l3 cache side-channel attack. In: 23rd USENIX security symposium (USENIX security 14), pp 719–732
16. Zhang Y, Juels A, Reiter MK, Ristenpart T (2014) Cross-tenant side-channel attacks in paas clouds. In: Proceedings of the 2014 ACM SIGSAC conference on computer and communications security. ACM, pp 990–1003
17. Zhang Y, Juels A, Reiter MK, Ristenpart T (2012) Cross-VM side channels and their use to extract private keys. In: Proceedings of the 2012 ACM conference on Computer and communications security. ACM, pp 305–316
18. Zhang Y, Li M, Bai K, Yu M, Zang W (2012) Incentive compatible moving target defense against VM-colocation attacks in clouds. In: IFIP international information security conference. Springer, pp 388–399
19. Ge Q, Yarom Y, Cock D, Heiser G (2018) A survey of microarchitectural timing attacks and countermeasures on contemporary hardware. J Cryptogr Eng 8(1):1–27
20. Heiser G (2021) How to steal encryption keys: your cloud is not as secure as you may think!. Microkerneldude
21. Herdrich A, Verplanke E, Autee P, Illikkal R, Gianos C, Singhal R, Iyer R (2016) Cache qos: from concept to reality in the intel registered xeon registered processor e5-2600 v3 product family. In: 2016 IEEE international symposium on high performance computer architecture (HPCA)
22. Lenoski DL, Weber W-D (2014) Scalable shared-memory multiprocessing. Elsevier
23. Lee Y, Kim J, Jang H, Yang H, Kim J, Jeong J, Lee JW (2015) A fully associative, tagless dram cache. In: ACM SIGARCH computer architecture news, vol 43. ACM, pp 211–222
24. Ahn J, Yoo S, Choi K (2014) Dasca: dead write prediction assisted sttram cache architecture. In: 2014 IEEE 20th international symposium on high performance computer architecture (HPCA). IEEE, pp 25–36
25. Sardashti S, Seznec A, Wood DA (2014) Skewed compressed caches. In: Proceedings of the 47th annual IEEE/ACM international symposium on microarchitecture. IEEE Computer Society, pp 331–342
26. Starke WJ, Stuecheli J, Daly DM, Dodson JS, Auernhammer F, Sagmeister PM, Guthrie GL, Marino CF, Siegel M, Blaner B (2015) The cache and memory subsystems of the IBM power8 processor. IBM J Res Dev 59(1):1–3
27. Yeh T-Y, Marr DT, Patt YN (1993) Increasing the instruction fetch rate via multiple branch prediction and a branch address cache. Int Conf Supercomput 93:67–76
28. González A, Aliagas C, Valero M (1995) A data cache with multiple caching strategies tuned to different types of locality. In: International conference on supercomputing. Citeseer, pp 338–347
29. Mancuso R, Dudko R, Betti E, Cesati M, Caccamo M, Pellizzoni R (2013) Real-time cache management framework for multi-core architectures. In: 2013 IEEE 19th real-time and embedded technology and applications symposium (RTAS). IEEE, pp 45–54
30. Liang Y, Mitra T (2010) Instruction cache locking using temporal reuse profile. In: Proceedings of the 47th design automation conference. ACM, pp 344–349
31. Altmeyer S, Maiza C, Reineke J (2010) Resilience analysis: tightening the crpd bound for set-associative caches. In: ACM sigplan notices, vol 45. ACM, pp 153–162
32. Vinothini S, Thirumalai CS, Vijayaragavan R, Senthil Kumar M (2015) A cubic based set associative cache encoded mapping. Int Res J Eng Technol (IRJET) 2(02)

33. Hellisp. CC BY-SA 3.0, An illustration of different ways in which memory locations can be cached by particular cache locations. http://creativecommons.org/licenses/by-sa/3.0/. via Wikimedia Commons
34. Microsoft (2017) Mitigate threats by using Windows 10 security features. https://docs.microsoft.com/en-us/windows/security/threat-protection/overview-of-threat-mitigations-in-windows-10. Accessed 30 Jan 2021
35. Percival C (2005) Cache missing for fun and profit
36. Irazoqui G, Inci MS, Eisenbarth T, Sunar B (2014) Wait a minute! a fast, cross-vm attack on aes. In: International workshop on recent advances in intrusion detection. Springer, pp 299–319
37. Inci MS, Gulmezoglu B, Irazoqui G, Eisenbarth T, Sunar B (2016) Cache attacks enable bulk key recovery on the cloud. In: International conference on cryptographic hardware and embedded systems. Springer, pp 368–388
38. Apecechea GI, Inci MS, Eisenbarth T, Sunar B (2014) Fine grain cross-vm attacks on xen and vmware are possible!. IACR Cryptol ePrint Arch 248
39. Schwarz M, Weiser S, Gruss D, Maurice C, Mangard S (2017) Malware guard extension: using sgx to conceal cache attacks. In: International conference on detection of intrusions and Malware, and vulnerability assessment. Springer, pp 3–24
40. Agrawal D, Baktir S, Karakoyunlu D, Rohatgi P, Sunar B (2007) Trojan detection using ic fingerprinting. In: 2007 IEEE symposium on security and privacy (SP'07). IEEE, pp 296–310
41. Page D (2002) Theoretical use of cache memory as a cryptanalytic side-channel. IACR Cryptol ePrint Arch 169
42. Jingfei K, Onur A, Jean-Pierre S, Huiyang Z (2012) Architecting against software cache-based side-channel attacks. IEEE Trans Comput 62(7):1276–1288
43. Neve M, Seifert J-P (2006) Advances on access-driven cache attacks on aes. In: International workshop on selected areas in cryptography. Springer, pp 147–162
44. Zhang T, Zhang Y, Lee RB (2016) Cloudradar: a real-time side-channel attack detection system in clouds. In: International symposium on research in attacks, intrusions, and defenses. Springer, pp 118–140
45. Gruss D, Lipp M, Schwarz M, Genkin D, Juffinger J, O'Connell S, Schoechl W, Yarom Y (2018) Another flip in the wall of rowhammer defenses. In: 2018 IEEE symposium on security and privacy (SP). IEEE, pp 245–261
46. Reza M, Amin H-F, Richard H, Farshad M, Mak S, Shahid S (2018) Are timing-based side-channel attacks feasible in shared, modern computing hardware? Int J Organiz Collect Intell (IJOCI) 8(2):32–59
47. Crane S, Homescu A, Brunthaler S, Larsen P, Franz M (2015) Thwarting cache side-channel attacks through dynamic software diversity. In: NDSS, pp 8–11
48. Yarom Y, Benger N (2014) Recovering openssl ecdsa nonces using the flush+ reload cache side-channel attack. IACR Cryptol ePrint Arch, p 140
49. Oren Y, Kemerlis VP, Sethumadhavan S, Keromytis AD (2015) The spy in the sandbox: practical cache attacks in javascript and their implications. In: Proceedings of the 22nd ACM SIGSAC conference on computer and communications security. ACM, pp 1406–1418
50. Weiß M, Heinz B, Stumpf F (2012) A cache timing attack on aes in virtualization environments. In: International conference on financial cryptography and data security. Springer, pp 314–328
51. Liu F, Ge Q, Yarom Y, Mckeen F, Rozas C, Heiser G, Lee RB (2016) Catalyst: defeating last-level cache side channel attacks in cloud computing. In: 2016 IEEE international symposium on high performance computer architecture (HPCA). IEEE, pp 406–418
52. Godfrey MM, Zulkernine M (2014) Preventing cache-based side-channel attacks in a cloud environment. IEEE Trans Cloud Comput 2(4):395–408
53. Varadarajan V, Ristenpart T, Swift M (2014) Scheduler-based defenses against cross-vm side-channels. In: 23rd USENIX security symposium (USENIX security 14), pp 687–702
54. Irazoqui G, Eisenbarth T, Sunar B (2016) Cross processor cache attacks. In: Proceedings of the 11th ACM on Asia conference on computer and communications security. ACM, pp 353–364
55. Gruss D, Spreitzer R, Mangard S (2015) Cache template attacks: automating attacks on inclusive last-level caches. In: 24th USENIX security symposium (USENIX security 15), pp 897–912

56. Pessl P, Gruss D, Maurice C, Schwarz M, Mangard S (2016) DRAMA: exploiting DRAM addressing for cross-cpu attacks. In: 25th USENIX security symposium (USENIX security 16), pp 565–581

57. Gruss D, Maurice C, Wagner K, Mangard S. Flush + flush: a fast and stealthy cache attack. In: International conference on detection of intrusions and Malware, and vulnerability assessment. Springer, pp 279–299

58. Lipp M, Gruss D, Spreitzer R, Maurice C, Mangard S (2016) Armageddon: cache attacks on mobile devices. In: 25th USENIX security symposium (USENIX security 16), pp 549–564

59. Bruinderink LG, Hülsing A, Lange T, Yarom Y (2016) Flush, gauss, and reload—A cache attack on the bliss lattice-based signature scheme. In: International conference on cryptographic hardware and embedded systems. Springer, pp 323–345

60. Backes M, Nürnberger S (2014) Oxymoron: making fine-grained memory randomization practical by allowing code sharing. In: 23rd USENIX security symposium (USENIX security 14), pp 433–447

61. Zhou Z, Reiter MK, Zhang Y (2016) A software approach to defeating side channels in last-level caches. In: Proceedings of the 2016 ACM SIGSAC conference on computer and communications security. ACM, pp 871–882

62. Evtyushkin D, Ponomarev D, Abu-Ghazaleh N (2016) Jump over aslr: attacking branch predictors to bypass aslr. In: The 49th annual IEEE/ACM international symposium on microarchitecture. IEEE Press, p 40

63. Gruss D, Maurice C, Mangard S (2016) Rowhammer. js: a remote software-induced fault attack in javascript. In: International conference on detection of intrusions and malware, and vulnerability assessment. Springer, pp 300–321

64. Gruss D, Maurice C, Fogh A, Lipp M, Mangard S (2016) Prefetch side-channel attacks: bypassing smap and kernel aslr. In: Proceedings of the 2016 ACM SIGSAC conference on computer and communications security. ACM, pp 368–379

65. Hund R, Willems C, Holz T (2013) Practical timing side channel attacks against kernel space aslr. In: 2013 IEEE symposium on security and privacy. IEEE, pp 191–205

66. Jang Y, Lee S, Kim T (2016) Breaking kernel address space layout randomization with intel tsx. In: Proceedings of the 2016 ACM SIGSAC conference on computer and communications security. ACM, pp 380–392

67. Roglia GF, Martignoni L, Paleari R, Bruschi D (2009) Surgically returning to randomized lib (c). In: 2009 annual computer security applications conference. IEEE, pp 60–69

68. Chirag M, Dhiren P, Bhavesh B, Hiren P, Avi P, Muttukrishnan R (2013) A survey of intrusion detection techniques in cloud. J Netw Comput Appl 36(1):42–57

69. John D, Matthew M, Jared S, Adrian T, Adam W, Simha S, Salvatore S (2013) On the feasibility of online malware detection with performance counters. ACM SIGARCH Comput Arch News 41(3):559–570

70. Keke G, Meikang Q, Lixin T, Yongxin Z (2016) Intrusion detection techniques for mobile cloud computing in heterogeneous 5g. Secur Commun Netw 9(16):3049–3058

71. Bhuyan MH, Bhattacharyya DK, Kalita JK (2013) Network anomaly detection: methods, systems and tools. IEEE Commun Surv Tutor 16(1):303–336

72. Gideon C, Hu J (2013) A semantic approach to host-based intrusion detection systems using contiguousand discontiguous system call patterns. IEEE Trans Comput 63(4):807–819

73. Weller-Fahy DJ, Borghetti BJ, Sodemann AA (2014) A survey of distance and similarity measures used within network intrusion anomaly detection. IEEE Commun Surv Tutor 17(1):70–91

74. Roberto P, Davide A, Prahlad F, Giorgio G, Wenke L (2009) Mcpad: a multiple classifier system for accurate payload-based anomaly detection. Comput Netw 53(6):864–881

75. Maurice C, Scouarnec NL, Neumann C, Heen O, Francillon A (2015) Reverse engineering intel last-level cache complex addressing using performance counters. In: International symposium on recent advances in intrusion detection. Springer, pp 48–65

76. Hu C, Liu J (2010) Method and apparatus for dynamic voltage and frequency scaling. US Patent 7,730,340

77. Tang A, Sethumadhavan S, Stolfo SJ (2014) Unsupervised anomaly-based malware detection using hardware features. In: International workshop on recent advances in intrusion detection. Springer, pp 109–129

78. Irazoqui G, Eisenbarth T, Sunar B (2015) S a: a shared cache attack that works across cores and defies vm sandboxing—And its application to aes. In: 2015 IEEE symposium on security and privacy. IEEE, pp 591–604

79. Liu F, Lee RB (2014) Random fill cache architecture. In: Proceedings of the 47th annual IEEE/ACM international symposium on microarchitecture. IEEE Computer Society, pp 203–215

80. Zhang Y, Reiter MK (2013) Düppel: retrofitting commodity operating systems to mitigate cache side channels in the cloud. In: Proceedings of the 2013 ACM SIGSAC conference on computer and communications security. ACM, pp 827–838

81. Seibert J, Okhravi H, Söderström E (2014) Information leaks without memory disclosures: Remote side channel attacks on diversified code. In: Proceedings of the 2014 ACM SIGSAC conference on computer and communications security. ACM, pp 54–65

82. Chen J, Venkataramani G (2014) Cc-hunter: uncovering covert timing channels on shared processor hardware. In: 2014 47th annual IEEE/ACM international symposium on microarchitecture. IEEE, pp 216–228

83. Wang Z, Lee RB (2007) New cache designs for thwarting software cache-based side channel attacks. ACM SIGARCH Comput Arch News 35(2):494–505

84. Page D (2005) Partitioned cache architecture as a side-channel defence mechanism

85. Kim T, Peinado M, Mainar-Ruiz G (2012) STEALTHMEM: system-level protection against cache-based side channel attacks in the cloud. In: Presented as part of the 21st USENIX security symposium (USENIX security 12), pp 189–204

86. Goran D, Boris K, Laurent M, Jan R (2015) Cacheaudit: a tool for the static analysis of cache side channels. ACM Trans Inf Syst Secur (TISSEC) 18(1):4

87. Leonid D, Aamer J, Jason L, Nael A-G, Dmitry P (2012) Non-monopolizable caches: low-complexity mitigation of cache side channel attacks. ACM Trans Arch Code Optim (TACO) 8(4):35

88. Bernstein DJ, Schwabe P (2018) New aes software speed records. In: International conference on cryptology in India. Springer, pp 322–336